D... KITCHENS

Children cooking. Postcard, 1903.

DOLL KITCHENS

1800-1980

Text and Format: Eva Stille
Photography: Severin Stille

Schiffer Publishing Ltd

1469 Morstein Road, West Chester, Pennsylvania 19380

Cover photo: Doll kitchen, circa
1910. Housing (36cm high,
59/42.5 wide, 30 deep) of wood
with white and blue tiling. Stove:
sheet metal, steel blue, nickel-
plated parts. Utensils: white
enamel, wood.
Back cover: pitcher with cup, circa
1800. Porcelain, green painted
trim. Height 6.6 cm. (Sig. G.
Ullmann, Munich)

Translated from German by Dr. Edward Force.

This work was originally published in German as *Puppenkuchen,
1800-1980* by Verlag Hans Carl Nürnberg

English edition copyright © 1988 by Schiffer Publishing Ltd.
Library of Congress Catalog Number: 88-614750.

Printed in the United States of America.
ISBN: 0-88740-138-4
Published by Schiffer Publishing Ltd.
1469 Morstein Road, West Chester, Pennsylvania 19380

This book may be purchased from the publisher.
Please include $2.00 postage.
Try your bookstore first.

Foreword

The subject of this book is the middle-class, predominantly German doll kitchen, which developed from a prestige and play item of adults to a toy for children toward the end of the Eighteenth Century.

Through the use of doll kitchens typical of their times, systematically collected over two decades, a continuing development up to the most recent past will be shown. The economic and social conditions that changed the real kitchens of adults also marked the doll kitchens. For that reason, each of the chapters about doll kitchens in various periods of time is preceded by a chapter about "big" real kitchens.

The particular charm of doll kitchens is in their richness of detail. Knives and forks, pots and pans of the most varied materials, small kitchen utensils and doll-house stoves are therefore treated very thoroughly. The relationship of children's household utensils to those of their mothers also helps to provide a better understanding of the subject.

From the great number of lovely old doll kitchens in museums—most of them already photographed often—only a few have been chosen deliberately, because the emphasis in this book, for several reasons, is on the period between 1870 and 1940. The most important changes in the household took place in this period, and in these decades more children owned doll kitchens than at any time before or since. And not least, kitchens from these times are still fairly available to collectors.

Acknowledgements

I would like to thank all the collectors, owners of toy catalogs, archives and museums—most of all Almut Junker of the Historical Museum in Frankfurt am Main—who generously lent me objects and private photos, made printed matter available or allowed it to be copied, and gave me important information.

At this time I also think gratefully of all those who parted with their toys for my benefit, or who, as dealers, for many years loyally made possible my access to "untouched" toys from private households.

Contents

Little Girls and Their Doll Kitchens

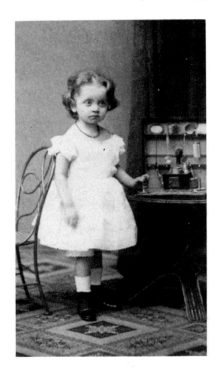

Girl with small doll kitchen. Visit-format photography, mid-Seventies. (E. Maas Photo Archives, Frankfurt/Main)

Dear Christ Child,

I would like a doll kitchen and a stove that I can poke up a bathtub and crayons 3 blocks and a kitchen cabinet and an ottoman.

Your Hilde

This is a translation of a five-year-old girl's 1939 Christmas list, written laboriously in German script (with a few words misspelled and a shortage of punctuation).

Doll kitchens, the dream of many girls, were a typical Christmas gift, a toy they could play with for only a small part of the year. When the supplies in the doll kitchen were used up, they cooked for days and weeks with supplies from the big kitchen, and their great inspiration gradually ebbed. And when the first beautiful days came and they could play outdoors, then, if not before, the doll kitchen season was over. The kitchen was carefully packed and put in the attic, and only in winter, shortly before Christmas, did the parents secretly bring it downstairs. Spruced up and filled again with goodies, it stood under the Christmas tree, bringing renewed excitement year after year.

When a girl outgrew her doll kitchen, then it was passed on to a younger sister, and then the firstborn girl of the next generation usually inherited the kitchen. Such traditions, kept up particularly in well-to-do families, can be observed in the example of a big cabinet doll kitchen whose history is as follows: This kitchen was commissioned in 1885 by a prominent citizen of the city of Bregenz—hotelkeeper, wine merchant and city councilman—for his oldest daughter Fini.

When Fini grew too old for dolls, the kitchen was given to her younger sister Marie, the great-grandmother of the present owner.—Marie later married and moved to Lorsch. Among her possessions when she married was the beloved doll kitchen of her childhood days. After her oldest daughter Helene, her younger daughter Elisabeth played with it from 1925 on. The two girls' father took care of the inherited doll kitchen like a "family relic". It was put away all year, but for every Christmas holiday it stood under the Christmas tree again, with highly polished copper and brass utensils, filled with sweets. Until the second of February (Mary's Candlemas) the children were allowed to play with it. That was the day when in many families the Christmas tree was traditionally taken down.

The kitchen then came to Frankfurt with Elisabeth's daughter Uli (born 1950). Today Uli's little daughter Luise, born in 1979, plays with Fini's doll kitchen.

During a hundred years of use as a toy, scarcely anything in this kitchen was changed. A few utensils were broken as it passed through the generations, and were replaced by new ones. But the back wall still is fully hung with old copper baking forms. And along with the black sheet-metal stove, the cocoa cans, copper pots, coffeepot and milk-warmer remained unharmed for four generations.

"Finchen's Doll Kitchen", 1885.
(Owner: U. Bähr, Frankfurt/Main

10

A Frankfurt doll kitchen was preserved just as lovingly, and a touching bit of history is linked with it. This kitchen is still decorated in the old way; even the crocheted shelf covers have survived through many decades of changing styles. Only the stove, for practical purposes, was replaced for every generation.

In 1933 Mrs. K. was given this doll kitchen. It had been offered for sale in a Frankfurt newspaper just before Christmas, because a Jewish family—about to emigrate to America—had to part with this relatively big doll kitchen. Three generations of women stood around it and cried as they said farewell to the favorite plaything of their childhood. Because of this history, Mrs. K. treated the kitchen with special respect from childhood on. She later set it up every Christmas for her daughter, and later for her granddaughter.

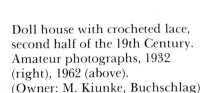

Doll house with crocheted lace, second half of the 19th Century. Amateur photographs, 1932 (right), 1962 (above). (Owner: M. Kiunke, Buchschlag)

Understandably, cooking was a special pleasure for girls and boys, more so than other kinds of housework. To be sure, the boys were only tolerated to a degree in this realm of the little housewives. The writer Marie Nathusius (born 1817) says this of a young friend named Alexander: "He often went along and was always welcome because he could play nicely with us, he poured the sugar, grated the chocolate and helped everybody."[1] And the writer and Egyptologist Georg Ebers (born 1837) remembers the role of the helpers in group cooking with the girls: It made us "happy when our sisters aloowed us to play the kitchen boys and helpers, in white aprons and caps."[2]

In addition, the boys must have enjoyed eating the more or less successfully cooked food. Berta D. (born 1908) says: "I had a farmhouse doll kitchen, painted blue, and under the table a chicken-coop with two chickens, and with lots of bowls and earthen pots and pitchers hanging up. We cooked by rubbing cookies and chocolate on a small grater and then mixing them with milk. My brothers ate everything up. Or we mixed up something for their menagerie. That was a big cage, divided up by bars to hold nine wild animals, who also needed food."

For Ludwig Ganghofer (born 1855), who loved a little girl who loved to cook, it was rather a sacrifice to eat the foods that were prepared: "I found Elsbethle... in the attic, where the child, anticipating my visit, had gotten out all the little kitchen utensils from the Christmas boxes. I didn't like to cook—that was "girls' work"—but I did everything for Elsbethle, even if I didn't like it. And whatever the quiet little cook produced, I swallowed without protest. But I was always happy when cooking was over and we sat down at the attic window or in a shadowy corner under the roof-beams."[3]

It was a question of ingredients, a matter of talent and a question of age too, as to what little girls put on the table. The very small ones always cooked inedible things. Marie Leske, a Nineteenth-Century author of books for young people, found chocolate soup made of water and black garden soil in the "Spielbuch für kleine Mädchen" ("Playbook for Little Girls", 1865), and coffee cake made of sand in "Vorstudien zur Kochkunst" ("Preliminary Studies of the Art of

Cooking"). She thought girls six to eight years old could "try their skill at really edible things", but only "cook cold". A really heatable stove could only be used at age ten, under supervision, and only at fourteen could girls cook alone. So all the good advice for relatively grown-up girls had been thought of and included in books about playtime and doll cooking.

The play gave way smoothly to the seriousness of real life. Appropriately, the introduction to "Haustöchterchens Kochschule für Spiel und Leben" ("Family Daughter's Cooking School for Play and Life", 1896) concludes: "So I wish you, my child, a pleasant success in all your undertakings with recipes old and new. Keep on growing, little daughter of the family, through happy play into serious life. What is a merry game for you now will still be a cherished activity for you later, and when one distant day it becomes your duty, it will be dear to you and you will playfully accomplish what you already practiced at play. Always cook tasty dishes for your dolls; you'll learn quickly, and you can please your dear Papa with his favorite dishes, splendidly prepared by your own hands. When it tastes good to him and he praises you, and you are your dear Mama's busy little household bee, then the grown-up daughter of the family will be as happy in her heart as the little mother of the dolls is today on Christmas Eve."[4]

The household was then the sphere of activity that most girls expected. Of those who could not attain the ideal of being a "housewife at her own hearth" many became servants, housemaids, housekeepers or cooks for strange "masters"; for unlike male servants, more and more of whom went off to work in industry, women worked predominantly in the household, even into the Twentieth Century. Of the 1,324,924 servants who worked in German households in 1880, 96.8% were still women.[5]

Doll kitchen, 1898. From an advertisement of the Mey & Edlich mail-order house, Leipzig-Plagwitz.

Adult kitchen, circa 1900.

Doll Kitchens—Mirrors of Their Times?

At first glance, the correspondence between "big" and "small" in the world of adults and the play world of children is astounding. This has led to the concept of toys expressed in slogans like "Mirror image of the adult world", and has given impetus to the attempt to use toys to look back on times past. That is correct under certain conditions.

In any case, the doll kitchen is only in part an accurate reflection of a big kitchen. Doll kitchens do not reproduce THE kitchen of a certain time, but ONE type of kitchen among many, and in fact that of the rather prosperous bourgeoisie, for only families at or above a certain standard of living could afford doll kitchens. But even when one has in mind this middle-class kitchen which was reproduced in the doll kitchens of the Nineteenth and Twentieth Centuries, one cannot expect a total agreement between "big" and "small". The room of a doll kitchen has only three walls, one for display and two sidewalls. The arrangement of the furniture and the stove does not follow the full-size pattern. It is arranged for the sake of the display wall's esthetics and to be reached easily by a child from the open side. The tendency to make all the nice things in the doll kitchen visible and reachable has, to formulate it a little too subtly, led to the Baroque "showy kitchen" being dragged into the Twentieth Century. While appliances and utensils had long since disappeared behind closed doors in adult kitchens, the walls of doll kitchens still hang full of baking forms and utensils, and the shelves are still full of plates.

The true-to-life nature of the little utensils often goes only as far as their appearance as well; that means that sometimes only some of the action takes place, because the real function no longer applies. For example, the handle of the bread-slicing machine can be turned, but the dull blade doesn't cut bread, the knife-sharpener doesn't sharpen knives and most coffee-grinders don't grind coffee because the beans are too big to fit through the tiny openings. All in all, many such machines and utensils are too small for "real" operation. Not by chance were there—and are there—bigger cook-stoves, appliances and utensils in addition to the doll kitchen furnishings, with which "real" cooking was practiced outside the doll kitchen. Cooking in the doll kitchen itself usually consisted only of so-called cold cooking of oatmeal, sugar and cocoa, that the child mixed thoroughly and in some cases stirred milk into. In the process a girl learned as little about cooking and keeping house as a boy learned about driving a car when he pushed a tinplate car through the room.

But the main thing, aside from the "learning to really cook" of the older girls, was probably the setting of the mood for a later role, the playtime activity with household appliances and machines, and the gradual gaining of familiarity with kitchen and household. And "playing as if" is quite sufficient for that; the similarity to Mother's kitchen is sufficient too, and that was always at hand. All the limitations that had to be made did not cut down the many actual similarities between "big" and "small". For example, the pieces of furniture in doll kitchens are amazingly like the big ones. And if one can become familiar with non-functioning machines, then one comes to understand just as many that do work. The girls could use simple devices like brooms, presses and mortars as well as more complex ones like grinders, coffee mills and butter machines. Of course little oil lamps really burned, and many little alarm clocks and hourglasses really could inspire one to wake up.

To understand the functional and stylistic developments of the doll kitchen as an idealized miniaturization of the kitchen, one needs to know the historical development of the economic and technical conditions in the kitchens of adults that provided the impetus for them.

The Kitchen with an Open Fire

Masonry stove with arch for firewood. Roasting turnspit and tripods. Taken from an illustrated pamphlet, "House and Kitchen Utensils", lithographed, beginning of the 19th Century. (Historical Museum, Frankfurt/Main)

The kitchen as a separate workroom, isolated from the rest of the dwelling, has only existed in Europe for a few centuries. Until the middle of the Sixteenth Century—and in rural areas much longer—the family life of a large household took place in one single central room, where the family was united not so much by blood ties as by common work and use of worldly goods.[6] The center of this room was the hearth with the open fire, from which they got light and warmth, and on which they cooked and roasted.

Even when people began to separate different rooms in the Seventeenth and Eighteenth Centuries, the room with the hearth still remained the central living and working area for a long time. Here the work involved in preparing food to eat was done, including what would later be located in specialized places outside the house: making butter and cheese, stuffing geese, raising poultry, butchering small animals, and baking bread. Characteristic of the old kitchen was the open fire. In the Low German language area the very low hearth was situated in the middle of the room, with a protective covering designed to stop sparks from flying, but without a chimney. The smoke rose through the framework of the roof, saturating it with tar, before it left the house.[7]

In the urban areas of North Germany the so-called pier-arch stove was also to be found, showing English and Dutch influence, characterized by closed sides, and set against a wall.

In South Germany too—the source of most doll kitchens—the stove was built against a wall quite early, because the room was heated by a stove located at the side of the room opposite the hearth.

The open fire was kindled on the masonry hearth until into the Nineteenth Century. In the old "fire

Open fire on a masonry stove
under a chimney flue.
Copperplate, 1854.

screen", or in the large flue that conducted the smoke out, the toothed kettle hook was attached to a crossbar. The kettle hung on it, over the open fire. For smaller pots and pans there were tripods and pan trestles to keep the necessary distance from the fire or flame. There were also small three-legged pots of clay or iron, as well as all kinds of roasting spits and warming pans.

The kitchen of that time was smoky and had dark walls, because the soot of the open fire fell everywhere. The floor consisted of stamped earth or was covered with large slabs of stone, or in some areas with square flagstones. For light there were pine torches, oil lamps, candles and tapers. Water was carried from the well. Many kitchens also had their own pump, as did, for example, the Goethe House in Frankfurt.

The kitchen utensils were in open cabinets, the dishes on plate shelves. Closing cabinets—where they existed—were used to store supplies in. In small coops poultry was kept alive until slaughtering.

"House and Kitchen Utensils".
From an illustrated pamphlet,
lithographed, beginning of the
19th Century.
(Historical Museum,
Frankfurt/Main)

"Open Fire" in the Doll Kitchen

In the Eighteenth Century the doll kitchen was already separated from the complex doll house. The oldest single doll kitchens of this era still show the typical features of the elegant Baroque doll houses, those status symbols of adults that served only marginally to educate children.[8] Only at the end of the Eighteenth Century did the doll kitchen achieve any noteworthy proliferation in the bourgeois class as a toy for children. That was not due to chance, for the economic and structural changes from the Sixteenth to the Eighteenth Century, which resulted in the development of large household families into small families, also brought a new concept of human beings. A person—and by extension also a "little person", a child—was now seen as an individual. A new understanding for childhood, which had not previously been regarded as a separate part of life, now developed, and brought, among other things, new pedagogical ideas with it. In this context toys were used not just for fun and pastime, but above all as means of education.

Doll kitchen, 1803.
(Hieronimus Bestelmeier No. 1012, Nuremberg)

Children's hearth for "open fire" with flue. D. Chodowiecki, latter half of the Eighteenth Century.

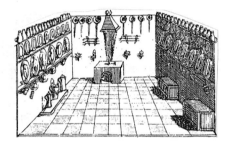

Doll kitchen, 1803. (Hieronimus Bestelmeier No. 400, Nuremberg)

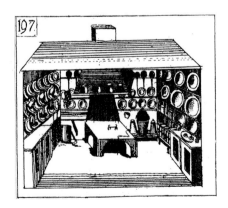

Doll kitchen, 1803. (Hieronimus Bestelmeier, No. 197, Nuremberg)

The most famous example of this is the toy catalog of the Nuremberg fancy-goods dealer Georg Hieronimus Bestelmeier. In the 1803 edition, along with "other useful things for instructive and pleasant activities of the young", very striking doll kitchen are shown in three different varieties:

"No. 400. A play-kitchen with furnishings whereof the utensils are of tin and porcelain. The kitchen in 14 inches long, 9 inches wide and 9 inches high, nicely painted, 3 fl. 48 kr. Box for it 24 kr." (Part II, Plate 8)

"No. 197. A new kind of play kitchen, it is much more natural than the formerly customary ones, in that it has a roof with a flue, a large door and two glass windows. Inside it is furnished very cleverly, one can push out the whole front wall, on the outside it is nicely painted, costs 4 fl. 30 kr., the box 36 kr. There are also very large ones available at 6 to 8 fl. If one wants the dishes, they must be paid for separately." (Part III, Plate 3)

"No. 1012. A play kitchen with running water, plus a cook such as come from the market in Nuremberg; the water tank is attached to the outside, and water runs into the kitchen through a brass tap, which can be turned on and off, into a trough. This gives children very much pleasure, also the kitchen is furnished with tin etc., 18 inches long and 12 inches wide and high, costs 8 fl. If such is wanted separately, the box costs 48 kr." (Part VIII, Plate 1)

In these Nuremberg Doll Kitchens, the side walls usually run together at the back, so that the kitchen was readily visible to the playing child and easily accessible everywhere. The hearth typically stood at the middle of the back wall. The arched opening under the surface provided a place to keep firewood. One had to imagine the "open fire" on the upper surface. The back wall under the big flue is appropriately black, sometimes even painted with tongues of flame.

Doll house with separate
cupboard, circa 1800.

The floor, painted in a checkerboard pattern,
imitates flagstones. The side walls—with several rows
of continuous plate shelves—are usually set with
wooden ridges on their upper edges to hold pots and
pans set on edge. Poultry pens and one or two closing
cabinets are usually built in. In addition to this basic
form, there are numerous variations. For example, the
Toy Museum of Nuremberg owns a diagonally shaped
one, and Bestelmeier offers No. 197 as a kitchen in
house form with moving wall and glass windows.

Probably from Augsburg. Housing (height with chimney, 65, width 87, depth 58 cm): painted wood, flue, shelves built in, water stand, hearth for "open fire"; tongues of flame on the back wall under the flue. Utensils of copper, brass, tin. Special features: fire shield to cover the flames (on the floor in front of the hearth). Doll circa 1830. (Doll museum, Wilhelmsbad, Hanau)

Small wooden doll kitchen.
(Nuremberg Pattern Books 1850-
60, page 37)

These "Nuremberg Doll Kitchens" were presumably made in large numbers. The other doll kitchens of this time were generally made individually by artisans as in previous decades. Museum pieces from local families can usually be identified clearly. For example, the Maximilian Museum in Augsburg can refer to many of its fine doll kitchens as "Augsburg Kitchens", while in museums in general the most commonly used term is simply "South German".

The contents of the kitchen, when one thinks in terms of earthenware utensils, also depends on the region. But the time is more decisive than the place of origin. In the first decades of the Nineteenth Century, the typical materials of the Eighteenth Century still prevailed: brass, copper and naturally wood. Bestelmeier of Nuremberg offered tin and even porcelain in his No. 400 doll kitchen as early as 1803.

A poem in the children's book "Hobby Horse and Doll", written by J. P. Wich of Nördlingen in 1847, lists:

What must be in a doll kitchen:
    Pans, tubs, buckets, vats,
    Canisters for salt and pepper,
    Knives, forks, spoons, pans,
    Pots, bottles, pitchers, cans,
    Wood and coal, fire tongs,
    Sulfur matches, flints, kindling,
    Mortars, oil and vinegar bottles,
    Roasting spits, roast turning jacks,
    Bellows, tripods, lids, sieves,
    Flour, meat, butter, grease and eggs,
    Plates, bowls, big and small,
    Must all be in the kitchen.

Along with the pans, tubs, buckets and vats for the water there were two water cans typical of Nuremberg, with which the water was "slid" or drawn from the well to the kitchen. Naturally, such "sliding cans" were to be found in many doll kitchens of the Nuremberg type.

Small sheet metal doll kitchen. (Nuremberg Pattern Books 1850-60, page 39)

Water container with two "sliding cans". From an illustrated pamphlet, lithographed, beginning of the 19th Century. (Historical Museum, Frankfurt/Main).

Interior of a Frankfurt kitchen,
circa 1850. Lithograph.
(Historical Museum,
Frankfurt/Main)

The Kitchen in the Mid-Nineteenth Century

In the first half of the Nineteenth Century, most kitchens remained as they had been. To be sure, the big households, in which family, workers and servants had been united under one roof, had been reduced further and were now found only among artisans and farmers. The production of much food had also been relocated outside the house, but bourgeois families still prepared some of what they needed at home. The chicken cages with living supplies of meat, and the chopping block on which small animals were slaughtered, butter churns, cabbage and pickle crocks speak for that. The sources of light were still simple oil lamps and candles. There was no running water yet. Water was still carried from the public well as before, unless the house had its own pump. Other than a few exceptional devices such as clockwork-driven roasting spits, the kitchen apparatus was still hand-operated. Things were made smaller with knives, graters, slicers and mortars. How much a kitchen of that time resembles an earlier one can be seen in a list included in the foreword of a book, "The Good Bourgeois Kitchen in all its Parts", by J. Rottenhöfer, Royal Cook and Master of the Household to Maximilian II of Bavaria:

It will not be purposeless if I indicate the equipment of a bourgeois kitchen here:

6 earthen meat pots of different sizes

6 ditto deep casserole-type bowls

6 ditto flat for baked goods of different sizes

2 ditto roasting pans

1 mixing pot of copper or brass

1 beating whisk of iron wire

1 ditto of wooden twigs for whipped cream

1 mortar of brass with pestle

2 baking pans

2 omelet pans of iron

2 butter forms of copper

2 creme forms
2 jelly forms
2 souffle/forms of copper
 or sheet metal
1 griddle
1 dumpling maker of
 white sheet metal
1 fish plate
1 small and 1 large
 skimming spoon
1 basting ladle
2 ladles
1 round pastry-cutter box
 with twelve pieces
1 ditto long—12 small
 ribbed biscuit forms
1 sugar shaker
1 scale with numerous
 weights
1 grater
1 funnel
3 brass pans
2 tart pans
12 individual tinplate
 pastry cutters
 comprising a set

2 pudding forms
1 anise-bread form (long
 rectangular sheet
 metal)
1 salt cellar
1 flour canister of wood
1 noodle board
1 rolling pin
1 cleaver
2 cutting boards
1 slicing knife
1 bacon knife
2 kitchen knives
1 cleaver
12 larding pins
1 trussing pin
1 strainer
1 chocolate machine for
 cooking chocolate with
 wooden stirring whisk
2 coffee grinders
12 cooking spoons
1 dumpling spoon of
 sheet metal
1 "ox-throat" cake form
1 water bucket

Stove from an adult kitchen, circa
1900.

2 water pitchers of
copper, tinned inside
several sheet metal
tablespoons
1 carving fork
1 kitchen lamp
1 tin freezing box with
wooden spatula
1 wooden bucket as well

1 sugar sieve
1 flour sieve
2 soup sieves
1 stuffing sieve
1 coal shovel
1 pair of fire tongs
1 brass doughnut wheel
1 cabbage slicer
1 pickle slicer.[9]

As old-fashioned as kitchen equipment of this time was, still work and life in the kitchen could be improved in one very important point: if a closed hearth were contrived. Such hearths, known as "thrift" or "saving hearths", had been developed late in the Eighteenth Century, but after centuries of being accustomed to an open fire, the change took place only slowly. So in spite of soot and smoke in the house, which according to a medieval proverb were one of the three worst plagues of household life (the other two were a leaking roof and an angry wife), the open fire remained customary until the middle of the century, even in the cities.

In rural areas people still cooked on an open hearth in the Twentieth Century, and in places in western Lower Saxony even after World War II![10]

The differences in the general course of life were formerly great. They depended on regional traditions, differences of rich and poor, city and country, and also on the conservative or progressive outlook of the population. All new advances—in housekeeping as well—existed for a long time side by side with the old ways. When they were finally established, they had already been superseded by something even more modern. The cooking place in the middle of the Nineteenth Century was therefore still an open fireplace on the floor in one house, on a small raised masonry hearth in another, while modern people already cooked on a hearth with an iron plate, or an additional high-walled baking oven, or even on one of the portable iron stoves, the so-called "cooking machines".

Sheet metal doll kitchen with
chicken cage, circa 1850. (Pattern
book of G. Striebel, State Archives,
Biberach an der Riss)

The walled tile stove enjoyed great popularity for a
long time, especially in South Germany and Switzerland. They were recommended to the "wives, daughters
and cooks" in Rottenhöfer's cookbook: "The best,
cleanest and most practical cooking stoves for every
household are definitely those stoves made by the potter
out of white glazed tiles, over which an iron plate lies,
which has round openings on top in which several iron
rings are set. Likewise a water container must be built
into these stoves, as also a roasting spit, which likewise
are heated by the same fire."

The Doll Kitchen in the Mid-Nineteenth Century

In the course of the first decades of the Nineteenth Century, more and more simple little doll kitchens came on the market, furnished only sparingly and equipped with only a few plates, pots and pitchers. They represented the modest outlook of the times, but their simplicity was also caused by their more and more widespread distribution in the lower middle class. The heavy tin containers and plates, the copper pots and kettles gave way gradually to the cheaper sheet metal painted to look like copper. There was more and more earthenware in these small kitchens, or utensils carved out of wood and painted so carefully and deceptively that one could take them for Fayence or porcelain if one knew them only from illustrations in old pattern books.

The housings of these simple little kitchens usually look just like the old doll kitchens of the Nuremberg type, with open-fire hearth under the big flue. At least

Doll kitchen in house form. Oskar Pletsch, circa 1850.

they are usually still shown in this form in pattern books and children's books from around the middle of the century.

But around this time, small iron stoves, the "cooking machines", are found in slowly growing numbers, in doll kitchens as well. They were often painted in tile patterns at that time, corresponding to the walled tile stoves in the big kitchen. The flue so typical of the open fire was still retained over the iron stove in the doll kitchen—as also in the big kitchen. Here, as in the big example, the stovepipe ran up into the flue.

At that time the small wooden hearth for "open fire" was often taken out of the doll kitchen and replaced by a modern sheet-metal stove. A characteristic example of the transition from "open fire" to closed stove in the doll kitchen can be seen on the title page of Julie Bimbach's doll cookbook, "Cookbook for the Doll Kitchen", which first appeared in Nuremberg in 1854: In front of a Nuremberg type of doll kitchen with its small masonry hearth for "open fire" a larger sheet-metal stove with stovepipe stands on a stool.

Doll utensils, turned wood, painted, 1850-60.
(Nuremberg Pattern Books, p. 67)

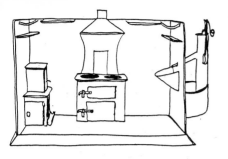

Sheet-metal kitchen, circa 1870.
(Original in the Munich State
Museum)

"Real" cooking on the cookstove first became possible on such an enclosed stove, for on the small wooden hearths whose stonework was just painted on, naturally no fire could have been lit, and only "cold cooking" could have been done. At that time there were iron stoves that were really stoked with wood, as well as those that burned alcohol. In 1865 Marie Leske wrote in her well-known activity book, "Illustrated Play Book for Girls": "Only when a little maiden is at least ten years old, may she cook on a really heatable stove, and then only under the supervision of Mama or a grown-up sister—for 'fire is a very dangerous thing, from which one cannot protect oneself enough'.

"A very capable girl of fourteen years may try her cooking skill alone with motherly permission, of course, but not on the alcohol flame. This cannot be warned about enough, because it spreads out around itself most easily." [11]

The joy of cooking took on a particular spirit when the implements were not tiny but corresponded to one's own size. Girls of the nobility enjoyed such privileges. "A kitchen into which one can walk, and in which the stove and pans are in proportion to one's own size, is an indescribable enchantment", wrote Carmen Sylva (Queen Elizabeth of Rumania, born 1843)[12], and Princess Marie of Erbach-Schönberg (born 1852) portrays a special surprise on her tenth birthday: "When I (got to the moss cottage), I discovered a wonderfully pretty, new-made stove very near it. Next to it stood Auguste Chapius in a white dress and held something covered in her hands. She congratulated me and asked me to take off the napkin. There on a wooden board lay tiny cutlets and sausages, just ready to be cooked! Then we went into the house. This was arranged like a farmhouse room, and with a closet that contained everything needed for housekeeping". [13]

"Nuremberg Doll Kitchen" under a Christmas tree, circa 1860. (Picture Stories for Small Children, Esslingen)

Sheet-metal doll stove in front of
an old "Nuremberg Doll Kitchen"
with wooden hearth for "open
fire". From: Julie Bimbach,
Cookbook for the Doll Kitchen,
Nuremberg, 1854. (Photo: W.
Schröder, Düsseldorf)

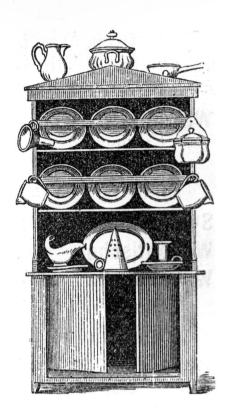

Stove for "closed-in fire", first half of the 19th Century.

The girls from "better" bourgeois circles (for whom the "Illustrated Play Book for Girls" was probably intended) also received Marie Leske's recommendation of a big stove. She advises them to write on their Christmas list:

1. A nice stove of cast iron, about one-sixth as big as our cookstove, in short, just right for your size, and made just like the one in the kitchen.

2. All the utensils that go with it, such as meat, soup and vegetable pots, water kettles, coffeepots, pans, bowls and saucepans, spoons of all kinds, tart and pudding and souffle/ forms, baking pans, sieves, potato mashers etc.

3. A well-filled food cupboard and

4. A sufficiently large supply of finely split wood.[14]

Sheet metal with brass oven doors, handmade; bronze pots, Nuremberg. Height 15 cm. (Munich State Museum)

"The City Kitchen", last third of
the 19th Century. (from
"Illustrated Instruction for
Youth", Esslingen, no date)

The Kitchen Between 1860 and 1900

Newly discovered or utilized sources of energy greatly lightened the burdens of housewives and servants in the second half of the Nineteenth Century. Gas, for example, which was first used in 1816 to light an industrial plant, was now used increasingly to illuminate the streets of large cities. Via the system of pipes created in the process, it could be piped into more and more factories, stores and finally private households. In 1868, 530 German cities were furnished with gas. The most important means of lighting at that time was nevertheless petroleum, which had replaced the oil lamp and the candle as an everyday source of light.

For cooking, of course, neither petroleum nor the alcohol that had already been used earlier was an alternative for wood and coal at first. The small cookstoves were only additional appliances and quick sources of heat for hours in which the hearth was not stoked. Once really perfected two-flame petroleum cookstoves were offered in the Mid-Eighties and gas had just been accepted as a source of energy for the household, electricity also began to compete around 1890. At first, of course, electric cooking was done only in well-to-do houses, for electricity was an expensive source of energy for a long time.

So the bourgeois kitchen, even in the last years of the Nineteenth Century, usually still had the big tiled masonry stove or the movable iron "cooking machine", or even the slow-combustion stove with a wood and coal fire, at its central point.

In the one-family house at that time, the kitchen was usually located in the basement and connected with the dining room upstairs by dumbwaiter and speaking tube. In tenement flats the kitchen was usually in a room facing the north, the dark inner courtyard or the shadowed side. Here too, the kitchen was separated

This is what an "electric kitchen" looked like in 1904. (Gehren, "Kitchen and Cellar", p. 90)

from the living rooms. Halls, doors and separate entrances simultaneously separated the servants from the family. The housewife of the upper bourgeoisie occupied herself only with overseeing the household, and when she could delegate this to a good housekeeper as well, she thoroughly enjoyed her representative duties.

In households of more modest means, that only had one "maid-of-all-work" and really could not even afford her, the servants had a particularly bad time. Thus the girls often changed their place of work, gradually going into factory work—for the sake of regular working hours—in preference to household work. Dissatisfaction increased on both sides and probably contributed to the entry of modern household appliances into the kitchen. On the one hand, it was hoped that they would attract good servants, and on the other, they made it easier to do without servants. This development increased toward the turn of the century, and as good as came to an end in the Twentieth Century.

The usual equipment of a kitchen of those times: the floor was covered with "Mettlach plates", hard bricks or terrazzo. Linoleum and wood floors also existed, but were regarded as unsuitable. The walls were usually painted with greenish oil paint. In one corner of the kitchen, usually by a window, the kitchen sink was built in. In kitchens that already had running water, the water faucet was naturally here too. But direct water connections were by no means to be taken for granted yet. At first, at the beginning of the Seventies, water pipes were laid only in the big cities. In apartment houses there was often just a single faucet on a landing between two floors for the use of several families. It was no wonder that the water bench, the "wet bench", on which the neatly covered bucket of fresh water stood, often continued to be used into the Twentieth Century.

Kitchen circa 1890 (anonymous photograph)

Kitchen and furnishings circa 1890. Gas stove at left next to the coal stove. Anonymous photograph. (Photo Archives of E. Maas, Frankfurt/Main)

The kitchen was furnished with at least one table with a hardwood top, a sideboard, several open shelves, a kitchen cabinet, several chairs, one of them unfolding into a stepladder, and a broom closet. A butcher block, a spoon box, a rack to hold cooking spoons, a lid holder and a cleaning box also belonged to the usual inventory. Naturally, the quality and quantity depended on the wealth and the population of the household. In better conditions, the utensils that had been found in well-equipped kitchens in the previous decades were enhanced with modern machines that made work easier, such as the steam cooking pots made by Papin and Umbach, refined French mechanical spits, graters, beaters, meat grinders, bean cutters, apple peelers, bread slicers and knife sharpeners, to name only a few.

In the decades before 1900, enameled utensils won a permanent place, and at the end of the century, aluminum utensils finally won their first friends in the kitchen.

The disappearance of the open fire generally created a friendlier atmosphere in the kitchen. This was emphasized by the fashion of hanging embroidered draperies and hand towels with cheery sayings, as well as putting embroidered lace on the edges of open and cabinet shelves.

The Doll Kitchen Between 1860 and 1900

In the latter half of the Nineteenth Century there were still, as before, small kitchens, usually broad in front and narrowing in back, with a stove in the middle and either racks on the sidewalls or the back wall covered with shelves or cabinets, leaving space for individual utensils on the sidewalls.

Another type of kitchen was common in these times: a very wide housing—120 cm is no rarity—approximately 45 cm high and 45 cm deep, sometimes on legs, sometimes with closing doors. A doll kitchen that took up space and presupposed a large dwelling. Its furnishings were splendid and rich.

In a city like Frankfurt with well-to-do bourgeois families, this broad type of kitchen with straight sidewalls enjoyed particular popularity.

Doll kitchen, circa 1850-60. Housing (height 46, width 91) of painted wood, red-and-white diamond-pattern floor, flue. Furniture: built-in shelves, painted brown. Water: reservoir with brass faucet and sink. Stove: sheet iron with brass doors and feet, tinplate cooking pots, brass knobs. Utensils: ceramic, stoneware, copper and tin. Special equipment: coffee grinder, coffeepot. (Historical Museum, Frankfurt/Main).

Most of these kitchens also had other characteristics in common: wall shelves on the left side and around the corner, with a corner cabinet under them, a sink with a drainboard in front, another set of wall shelves on the right, with a bench or small cabinet underneath. A kitchen cabinet on the rear wall at right, and the iron stove in the middle, usually under a flue. The floor in a checkerboard pattern, walls and furniture in subdued colors, often finished in beer glaze.

These kitchens' similarity to each other led to their being called "Frankfurt Doll Kitchens", and to the assumption that they were made by a Frankfurt artisan.[15]

The rich furnishing of these and other doll kitchens of the time can be seen in a "Small Illustrated Home and Household Book for our Darlings". Dressed up as a story, "as an inducement to independent thought and creativity in the household sense. Prepared according to attainable principles and published by Elly Gregor and Johanna von Sydow," with the title "Lizzie's Doll Kitchen" (Lieschens Puppenstube), it first appeared in Leipzig in 1884:

"Lizzie knew exactly what belongs in complete kitchen and house furnishings:

Lizzie's Doll Kitchen, 1884.

Sink with drainboard, water
bucket and wash basin. (for the
kitchen beside it, 1860-70)

A kitchen cabinet to protect many things from dust; a
washing table and besides that a big and comfortable
kitchen table, as well as one or two kitchen chairs. Also
a fly cupboard is very useful, so that the various tasty
foods are protected from nibbling flies; several
casseroles, pans and stewpots, two or three saucepans
and just as many roasting pans—Lieschen has baked in
them like so many a little one—a coffeepot, yes, because
it is very important to have a special cooking vessel for
the beautiful brown drink; one certainly can't make it
in the bouillon pot; the coffee tastes best only when it is
brewed well. Then: a small kettle for stewing fruit and
mixing purees. . .

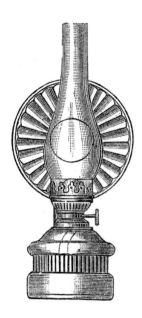

Petroleum lamp for a doll kitchen, circa 1880.

Canister, circa 1890. Sheet metal, lacquered and printed with an onion pattern.

In the kitchen there should also be an egg poacher in which each egg has its own small cooking pan, a pancake turner, so that the round, thin cake can be turned easily; also a coffee weight, a whole, half-, quarter- and eighth-liter measure, as well as a scale and smaller weights, for if the foods are to turn out right, the ingredients must be measured well; then a lid holder with pot lids of sheet metal, several sheet-metal bowls, a wash basin to hold water; a skimming ladle, a bouillon sieve and a strainer, so that everything can be removed from a broth that mustn't go into the dish; a coarse and a fine grater—do be careful that your fingertips don't get rubbed bloody! several funnels, a milk can that has to be scalded out well after every use; a tea filter, several forms for baked goods—for baked goods cook, or rather bake, only too willingly for little mouths that enjoy their taste—a fomenting form for ash cake, cake pans, cookie cutters. The last are especially necessary to make small ornaments for the Christmas tree; an ice cream form is also handy to have; with every better dinner Lieschen served well-made ice cream. A dozen ragout pans also belong in a finer kitchen, and a little girl who knows how to make a good-tasting ragout is already considered an accomplished cook. Ladles, vegetable spoons and ordinary sheet-metal spoons must be there in sufficient numbers, for one cannot possibly stir sour and sweet foods, bouillon, milk or the like with one and the same spoon; a beater for egg-white, a kitchen wheel to cut dough, a brass mortar, a slicing knife and a cleaver, a carving knife, a few butter and cheese knives, strong and fine skewers, a knife-sharpener, a corkscrew, a dozen table knives and forks, a dozen dessert knives and forks, a dozen tablespoons and just as many teaspoons, a dozen knife-rests, a tablecloth, so the rather hot vessels don't spoil the polish of the table, a half-dozen wooden ham platters, a salad set of horn, mustard spoon, half a dozen horn vegetable knives, pepper and salt shakers, an egg timer (sandglass), a dozen flat porcelain plates and a dozen soup-bowls, as well as a dozen dessert plates, long and round roasting pans, saucepans, round and square, flat and deep bowls and vegetable pans of different sizes, two soup tureens, a potato pan with a lid, coffee- and teapots, cream pitchers, a sugarbowl, big and small kitchen plates, water bottles,

Picture from a paper-pasting
game. "The Dollhouse", Mainz,
circa 1875. (Sig. Köstler, Munich)

glasses for water, wine and liqueur, a carafe—all
according to your needs; the more prominent the
household and the more guests present, the more
inclusive the kitchen furnishings must be. And don't
forget cutting boards, a cooking-spoon rack with
spoons and beaters, a meat mallet, a rolling pin, salt
and flour canisters, egg boxes, so the fragile things
don't fall to the floor, a spice rack, curling irons, lamp
shears, a hot-water bottle, flatirons with a rack, kitchen
lamps and lanterns, a spirit lamp with boiler, a
breadbox, potato peeler, bronzed coffee, sugar and tea
containers, a sprinkling pot, a crumb scoop with a
small brush, a breadbox to keep the bread fresh; also a
pestle, roasting racks to keep the roast from burning, a
tripod, a coffee mill, a noodle board, a pickle slicer, a
fishnet, a waffle iron, a bottle basket, a knife-cleaning
board, scouring powder, a cleaning box, a wax box and
a sand box, various brooms, brushes, a dustpan—sweep
gently and softly so the dust doesn't fly around!—a
washcloth basket, a curtain brush, clothing, velvet,
shoe and scrubbing brushes, bottle and plate brushes, a

Iron stove with copper and brass,
handmade, copper pots, circa
1890.

cylinder cleaner, a rug beater, cleaning cloths,
dustcloths and pot cloths; also water pails and water
pitchers, dishpans, wash basins, soap dishes, washlines
and clothespins, laundry, carrying and wood baskets, a
woven hamper for dirty clothes, market basket, coal
scuttle, coal shovel and fire iron, a kitchen fire lighter,
ashcan etc.''[16]

In the doll kitchens of the Nineties there were also all
kinds of small machines such as bread slicers, ice cream
machines, butter machines of glass with churning
mechanisms, grinders, milk warmers, petroleum lamps
and finally small iceboxes; pantries were by no means
rare.

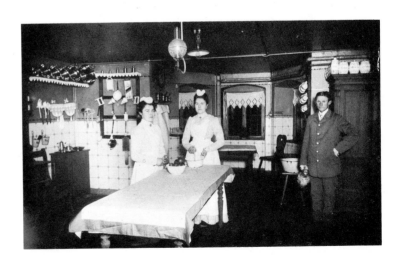

Kitchen with furnishings circa
1880. Interesting details from left
to right: brass kettle, onion nets,
porcelain utensils, mortar,
porcelain canisters, lace curtains.
(Photo: G. Sauer, Wittenberg,
circa 1890)

Doll kitchen, circa 1900. Housing (height 44, width 81.5, depth 45.5 cm) of lacquered wood, greenish tile pattern on the floor.
Furniture: old style, modern color wood, lacquered white. Water: sink. Stove: sheet iron, patterned, brass doors and feet. Enameled cooking pots (see color plate). Utensils: porcelain, stoneware etc. Special equipment: butter machine.

Doll kitchen, circa 1905. Housing (height 41, width 99, depth 45 cm) of lacquered wood, base with tile-pattern covering. Furniture: old style, modern color: white and blue lacquered wood. Water: sink. Stove: sheet steel, with bricklike pattern, tinplate cooking pots. Utensils: ceramic, porcelain. Special equipment: icebox (at right). (Historical Museum, Frankfurt/Main)

The Bright Kitchen, 1900-1920

Dark, insensitive colors had been necessary for the kitchen as long as smoke and soot from the open fire constantly rained down on walls and furniture. People had gotten used to it, and even in the last decades of the Nineteenth Century, when the closed stove had become the usual thing, greenish-gray painted walls were still preferred. But at the end of the century the kitchen caught the attention of the hygienists, the economic theorists, architects and artists. At that time the first step was made toward a whole new conception of the kitchen.

The modern kitchen should be a bright, airy room, with several subsidiary rooms when possible. Hygiene was taken almost as seriously in the kitchen as in the hospital. Runing water should be taken for granted.

Joseph August Lux, who popularized the ideas of his contemporary artists in various writings, described the ideal kitchen in his 1905 book, "The Modern Dwelling and its Furnishing":

One "makes the windows wide and fairly high . . . , so that the wall surfaces can be put to good use in the kitchen layout. Under these windows there are usually cabinets with as many shelves and storage places as possible, closed by glass doors. In the middle of the wall, under the window, we generally find the worktable; its lower parts are used as a cabinet, and it is crowned by a moulding with closing compartments. On the opposite side stands the stove. Unlike the kitchen of the past, which was regarded as beautiful only when the shining brass and copper pots, the colorful bowls of stoneware and porcelain, the tin and sheet metal containers were impressively displayed on walls and open places to delight the housewife, today people love to shut away every kitchen utensil in the cabinets, and are completely right to do so . . . For thus can the utensils be kept free of dust and flyspecks, and one is spared a lot of cleaning work . . . Only the copper pots are left hanging in the open. But such a kitchen looks appetizing enough, namely when the walls are tiled in white, as is often the case of late . . . In place of tiles, thin marble slabs are also used, and of course only white ones, because for understandable reasons it is a principle that white should predominate. Therefore all the wooden objects, all the kitchen furnishings, are lacquered white, which gives the advantage of making any dirt easy to remove by simply washing it off. The fact that one sees any uncleanliness at once on white is only an advantage, because it should never be tolerated anywhere, and least of all in the kitchen. If one wants an ornament, it should be a flat ornament, smoothly patterned and used sparingly . . .

"The kitchen furniture should have surfaces running down to the floor and stand firmly on it without legs."[17]

This concept, hygienically white, more laboratory than lived-in kitchen, was not agreed to by all his contemporaries. Rudolf Mehringer, for example, looked with regret at the increasing far-reaching changes in the kitchen. In 1906 he wrote in his book, "The German House and its Furnishings": "The kitchen has experienced a strange fate. Once the midpoint of the house, the housewife's place, it is regarded today as more and more of a burden and pushed away from the living rooms. The woman of the house withdraws from it more and more."[18]

To be precise, the wishes or, as the case may be, the fears conceived of by the people of those days were realized only in the Fifties and Sixties of our century. Around 1900, the modern, electrified laboratory type or the esthetic Youth Style kitchen was still opposed by the old-fashioned kitchen of the Nineteenth Century.

Color plate
Kitchen utensils large and small.
Upper left: water kettles, circa 1925, aluminum. Height 15 and 7.3 cm.
Upper middle: storage canisters, circa 1920-30, stoneware and porcelain with wooden tops. Height 18.5, 15, 5.3 and 4 cm.
Upper right: oil cans, circa 1900, red-painted sheet metal, brass screw caps. Height 26 and 6.2 cm. (Sig. G. Ullmann, Munich)
Center left: butter machine, circa 1890, glass with cast iron parts. Height 40 and 13.7 cm.
Center middle: kitchen oil lamps, circa 1900, glass and brass. Height 25 and 8.5 cm.
Center right: pots of a "food carrier", circa 1900, stoneware and porcelain. Height 10 and 3.2 cm.
Lower left: coffeepot, circa 1880, porcelain, blue finish, probably from Huttensteinach, Thuringia. Height 21 and 9.5 cm.
Lower middle: Bread-slicing machines, 1930»s, cast iron on wood, painted red. Height 18 and 5.5 cm.
Lower right: cooking pots, circa 1800, wrought iron, wooden handles. Height 23 and 10 cm. (Historical Museum, Frankfurt/Main)

Youth Style kitchen, designed by Patriz Huber, Darmstadt, 1900.

Bright kitchen, 1905-10. Light
tiled walls, similar tiled stove,
white porcelain canisters and
pitchers. Amateur photograph.
(Photo Archives of E. Maas,
Frankfurt/Main)

Kitchen, 1901.

The buffets with their turned pillars and intricate woodwork stayed until into the Twenties. Palace-like kitchen cabinets with projecting and even bowed middle sections, latticework around the etched or polished windows, machine-cut ornamentation in the door panels seemed "modern" to most housewives. White lacquered surfaces set off with black decorative designs and stained natural wood (light to dark mahogany) were preferred then.

"Nuremberg Doll Kitchen", circa 1800. Housing (39.5 cm high, 73 to 52.5 cm wide, 38 cm deep) of painted wood. Furnishings built in. Hearth for "open fire" wth flue. Utensils of copper, tin and wood. Special features: brass utensils from Eighteenth Century, (some added later).
(Historical Museum, Frankfurt/Main)

Plates for the table, circa 1850. Turned wood, painted. Height of the tureen 3.2 cm.

The Bright Doll Kitchen, 1900-1920

Even more than the big kitchen, the doll kitchen retained many elements from the Nineteenth Century for a long time. Most of all, one did not want the pretty utensils, the pots and implements to disappear into closed cabinets, as would have corresponded to the demands of the time. But in most points the formation and furnishing nevertheless thoroughly represented the spirit of the times.

The bright hues, natural wood finishes, beige, ivory, white with light blue or bronze lines became very popular; even the sparingly used, flat-patterned ornament was applied gladly. For doll kitchens in these times, "tiled" walls were typical, either painted or papered, often in Dutch patterns. A toy dealer in Baden-Baden offered corresponding kitchens in a 1913 catalog: "straight-lined, lacquered in white and blue" and "four-cornered, diagonal, with water pipes and faucet, imitated Majolika wall design (blue ornaments on white porcelain background), opening windows and sheet-metal stove with operating utensils."

In the first two decades of our century there were doll kitchens in all sizes, according to demand and space conditions, as well as financial considerations, from the tiny "Erzgebirge Kitchen" via the "assemblable", the "small wooden decorated with pyrography", and the very flat "French Kitchen", to the "very large kitchen" with "palacelike" cupboards. The attitude of adults to toys had changed somewhat at this time, the esthetic pleasure and the personal inclination of the child were given more consideration now.

Paul Hildebrandt's inclusive book, "The Toy in the Child's Life", which appeared in Berlin in 1904, lets these aspects shine through in his portrayal of "doll-kitchen splendor":

Water bench with bucket, pitcher, console and wash basin. (see the color plate opposite).

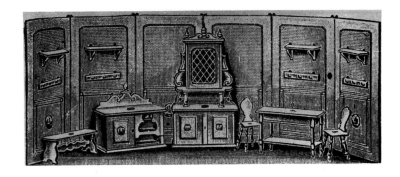

Assemblable kitchen, 1903-04, offered in a department store catalog. (Possibly made by the Märklin Brothers, Göppingen)

Doll cookstove, circa 1910. Sheet iron, nickel-plated parts and utensils. Height without stovepipe 10 cm. (Historical Museum, Frankfurt/Main)

"For a girl who has the spirit and desire for household work, and who belongs to the type of children in whom the talent for handiwork is stronger than purely intellectual tendencies, such a big children's cookstove or even a kitchen cabinet with assemblable cookstove and wall construction for the kitchen implements is one of the best and most useful presents. Such a present can also be called an artistic one from the modern standpoint, in Van de Velde's sense, without doubt. How friendly and bright is the pretty kitchen cabinet painted with white Japanese lacquer, with its blue lines on the cabinet doors and the red tile lines on the wallboards. On the cabinet, which is as high as a child's table, stands the artistic, elegant cookstove with its glowing golden yellow protective rod (through which the gas is conducted when cooking with gas) and the brass doors splendidly set off from the dark iron body."

Doll kitchen, 1910. Housing (height 8, width 22 tapering to 19.7, depth 9.2 cm) of wood. Lacquered furniture. From the Erzgebirge.

Kitchen furniture with food, circa
1860. Furniture of painted wood.
Food on cardboard plates is made
of tragacanth and paste. Height of
the cupboard 35 cm. (Historical
Museum, Frankfurt/Main)

Above: doll kitchen, 1860-70,
housing (height 48, width 120,
depth 55 cm) of painted wood,
floor originally white, red
checkerboard pattern, flue.
Furniture: painted wood, built-in
shelves. Water: water tank with
faucet, sink. Stove: sheet iron with
brass, for alcohol; tinplate cooking
pots and brass kettle. Utensils:
stoneware, porcelain, earthenware,
pottery, sheet metal and copper.
Special equipment: food carrier.
(Historical Museum,
Frankfurt/Main)

Below: doll kitchen, circa 1875.
Housing (height 41, width 112,
depth 52 cm) of painted wood.
Furniture of painted wood, built-
in shelves. Water: large water tank
with faucet and sink. Stove: iron,
massive, brass details, made by an
artisan; cobalt blue enamel
cooking pots. Utensils: porcelain,
blue underglaze, strawflower
pattern, wood and wire goods.

Doll kitchen under a Christmas
tree, 1910-20. Amateur
photograph.

"The plate of the cookstove is made of mirrorlike
sheet iron, and the cooking pots of nickel-plated zinc;
how brightly, how amiably and neatly they glow at the
little recipient, as if they wanted to say: 'We'd like to be
kept so shining clean by you too', and what versatility
is discovered in them.

Doll kitchen, 1910. Housing (height 28, width 59 tapering to 47, depth 28 cm) of lacquered wood, base and floor papered in tile pattern. Water: sink. Stove: sheet iron with doors, feet and frame of brass-colored sheet metal, for spirit, tinplate cooking pots with porcelain knobs. Utensils: porcelain, wood. Special equipment: lid holder with lids, wall shelves with carving set and whetting steel (in back), sheet-metal ladder chair at right.

"There we see a tea- or water-kettle, casseroles, cooking and stewing pots, fish cooker, coffee machine, chocolate cooker and the baking pans for two present baking and roasting ovens; with a water tank for hot water and a stovepipe this splendid cookstove, which can be heated with either alcohol or gas, is naturally also supplied.

Doll kitchen, 1910-20. Housing (height 28, width 67 tapering to 56, depth 36.5 cm) of lacquered wood, base light blue and white, floor brown, black and white. Furniture: lacquered white, adhesive pictures. Water: reservoir with faucet and basin. Utensils: porcelain, sheet metal, painted light blue. Special equipment: sheet metal on the wall printed in light blue, with scrubbing brushes, plus wash basins.

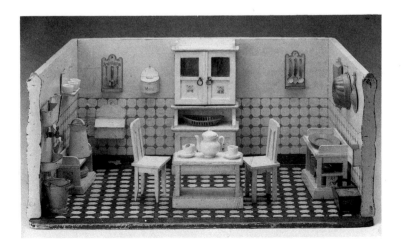

Above: doll kitchen, circa 1860-70. Housing (height 35.5, width 81, depth 40 cm) of wood, beer glaze. Water: sink, washbowl, container for fresh water near the stove. Stove: iron, massive, for alcohol, handmade; copper cooking pots. Utensils: tin, porcelain, earthen pots. Special equipment: coffeepot on the stove, coffee mill on the table at right, oil lamp on the bench.

Below: doll kitchen, circa 1875. Housing (height 19.5, width 55, depth 34 cm) of wood, intarsia. Furniture: natural wood. Stove: sheet iron, brass-colored doors and lion's feet, for alcohol; tinplate cooking pots with brass lids. Utensils: ceramic from Bunzlau. Brass kettle. Special equipment: spice rack, butter tub.

Above: doll kitchen, 1860-70. Housing (height 50, width 118, depth 51 cm) of painted wood. Furniture: wood, beer glaze. Water: sink, water bucket, wash basin. Stove: sheet iron; doors, corners and feet of brass, handmade; cooking pots of tinplate with brass lids. Utensils: earthen pots, sheet metal tableware. Special equipment: vinegar barrel.
Below: doll kitchen, 1880-90. Housing (height 42.5, width 100, depth 42.5 cm) of painted wood, wall repainted circa 1900. Furniture: wood, beer glaze. Water: sink, bucket. Stove: sheet iron, marked to look like tile, doors and lion's feet of brass; tinplate cooking pots. Utensils: ceramic, porcelain. Special equipment: onion net, lid holder.

Doll kitchen, circa 1900. Housing (height 25.5, width 41.5, depth 40.5 cm) of wood with tile papering, walls in Dutch pattern, window. Furniture: Natural wood with pyrography. Stove: with back wall, stamped sheet metal, brass-colored door, tinplate cooking pots with sheet-metal lids. Utensils: wood and porcelain.

"But with the description of the cookstove we have come nowhere near exhausting the splendors that such a children's kitchen contains, for everything found in our real kitchen is also at hand in the children's kitchen.

Such as those yellow-lacquered canisters for the storage of coffee, sugar, lard, flour, then the salt can, the little containers for the spices, plus a grater, sieve, funnel, cake, food and pudding forms, cleaver and slicing knife, various carving knives, coffee mills, ice cream, butter and breadcrumb machines, kitchen scales, pestles, strainers, and finally buckets, pans, dishpans, as well as all the porcelain, sheet metal and enamel utensils that belong to cooking.

"All these utensils are housed in the kitchen cabinet and on the kitchen shelves and frames, but we also find in the complete children's kitchen a whole array of kitchen furnishings and devices that are not missing in a good household.

"These include the kitchen table, kitchen chairs, icebox, and the broom rack complete with broom, scrubber, garbage pail, brush, dustpan and rug beater.

Even the shoeshine box is there with all its equipment."[19])

Doll stove for the kitchen shown at top on page 89.

Doll stove for the kitchen shown at bottom on page 89.

The Reform Kitchen of the Twenties and Thirties

During the first World War and in the years after it, the self-image of women changed significantly. The independence they achieved and their experiences in these times, along with the more and more customary small family and decreasing numbers of children, furthered women's occupational activity and finally led to a new self-awareness. Women no longer saw their most important task in housekeeping, but rather a job that had to stay on the fringe of their lives. This was possible only because they began to regard the household as being like an independent small business—at least in terms of getting work done—and therefore make it thoroughly practical, electrified and mechanized. The layout of the kitchen was now planned carefully, every step and every use of a hand considered. In the kitchen, as in workplace analysis, the shortest way for every activity was found and the simplest method worked out through calculation and motion study. All work that could be done by machines should be done by them, if possible by electric machines. Bit by bit, electricity was to replace muscle power in the kitchen. That was the ideal concept of the time.

The housing shortage of the Twenties helped to advance the concept of a small and practical kitchen. Social Democratic city governments and architects with a sense of social responsibility set up housing projects based on the minimal incomes of the tenants. In the big cities, housing developments were built that sought to offer the greatest possible quality of life in the smallest space. For example, between 1923 and 1926 Vienna built 25,000 apartments that included only 38 square meters of floor space each and yet were arranged to be optimally useful for four persons (parents with

Color Plate
Doll kitchen, circa 1890. Housing (height 45, width 79.5, depth 47 cm) of painted wood, with two windows. Furniture: wood and sheet metal (hanging shelves), lacquered. Water: sink, water bucket. Stove: sheet iron with copper, handmade by an artisan; copper cooking pots. Utensils: tin items, brass kettle. Special equipment: bread slicer, lace curtains.

"Frankfurt Kitchen", 1926
(Designer: Grete Schütte-Lihotzky,
Frankfurt/Main)

Reform Kitchen, 1930.

In the living roomkitchen, 1930.
Amateur photograph.

two small children). A practically appointed cooking niche with roller blinds, removable and folding components replaced the kitchen here and used up only 2.3 square meters.[20]

The pattern for many later designs of modern kitchens was the so-called "Frankfurt Kitchen" created by the Frankfurt architect Grete Schütte-Lihotzky in 1926; it was also known as the "May Kitchen" because it was built into series of "May Developments". This kitchen, in terms of its division of space, inclusion of built-in cupboards and storage facilities and totally planned development, resembles the kitchen of a railroad dining car.

This development was actually inspired as early as 1869 in the ideas of the American, Catherine E. Beecher.[21] She already recognized then that the work of a housewife could be reduced by a rational arrangement of the kitchen and its work processes. In detailed descriptions and drawings she followed a similar course to that of Schütte-Lihotzky: in place of the kitchen table she planned extensive work areas along the walls; instead of the cabinet, various wall shelves and drawers or containers under the counters, so that all utensils were ready to be picked up on the spot where they were needed. The organizing of work processes thus predates by many years their mechanization. The actual mechanization, which means electrification, can be regarded as complete in Germany only after World War II, of course, but even in the Twenties and Thirties it was expanded intensively.

This perfctly rational type of kitchen was, as said, built into series of housing projects. Architects designed larger but in principle similar kitchens for themselves and prosperous clients, but the public had scarcely begun to accept these purposeful, coolly esthetic kitchens at that time. Kitchens without tables, without chairs, intended only for the housewife who was able to do the most necessary kitchen work there alone and as quickly as possible, did not seem to represent the needs of most people, who obviously preferred a "cozy atmosphere" and simply wanted to "be there". Amateur photos and kitchen furniture catalogs of the Twenties and Thirties, in any case, show far fewer rational arrangements than lived-in kitchens.

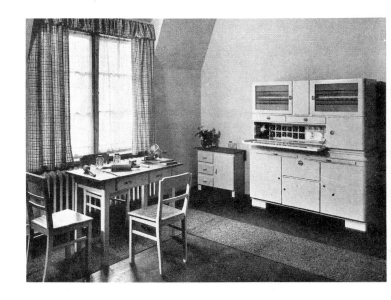

Reform Kitchen, 1937.

The kitchen cabinets show decorative moldings, channeled and ridged, windows with curved or gridlike wooden mouldings or etched patterns, and bowed fronts with strongly veined panels of natural wood. To some extent, ivory or lime-green lacquered closets with mottled wood moulding were preferred, on broad, curving feet.

The utilized variation of the uncompromisingly rational, purposeful architect-designed kitchens of the time was the so-called "Reform Kitchen". Characteristic of this kitchen type is a wide, low kitchen cabinet with smooth front and built-in storage bins, a comfortable swivel chair, a table with drawers, perhaps a dishwashing table or counter by the sink. All wooden parts, finished in white or ivory lacquer, were easy to care for.

This "Reform Kitchen" was accepted as the ideal into the late Thirties and was revived around 1950. In both eras it was also the model for doll kitchens (see illustrations, pp. 75 & 95)

Doll Kitchens of the Twenties and Thirties

As in big kitchens, the modern, rational kitchen was naturally something special in doll kitchens too, perhaps even more so. One of the doll kitchens shown here, housed in a flat closing case taking up the least amount of space, has (exceptionally) the most important requisites of a modern kitchen of the Twenties: gas stove, hot water heater, smooth fronts. Everything arranged ready to use or at least near the workplace. But in general, almost all toy kitchens were arranged in the old way, with individual pieces of

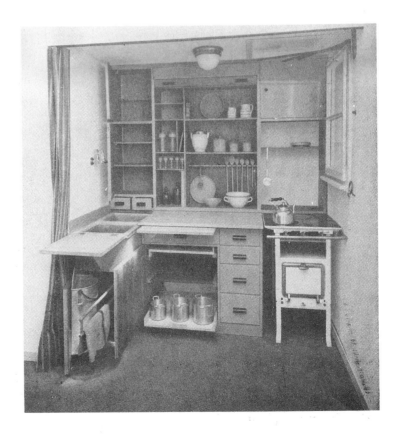

Cooking niche, 1925. (Franz Schuster, Vienna)

Doll kitchen, Twenties. Amateur photograph. (Photo Archives of E. Maas, Frankfurt/Main)

Doll kitchen (and grocery store), Twenties. Amateur photograph.

Doll kitchen cabinet, 1910-20. Natural wood, mahogany stain. Etched windows with black frames, counter covered with green linoleum. Height 44 cm. (Owned by K. Krier, Frankfurt)

furniture, a "real" kitchen cabinet, table, chairs and many things on the walls. Perhaps not just because the child could play with them better, but also because the mother was emotionally linked to the "cozy" kitchen of her childhood and chose or furnished the doll kitchen accordingly from this standpoint:

"Our girl was in an even more excited mood than the others—if that was possible. A doll kitchen awaited her on the present table, one long dreamed of and wished for! I really must describe it. You will see it, of course, but just let me tell you what we made out of Aunt Emma's old doll kitchen: Well—a brightly papered kitchen with a real, genuine window to open and close, on which even the airy curtains can be drawn; furnished with a flowered ottoman, chair, table, bench, big opening kitchen cabinet (=closet) and small sideboard, all in lime-green lacquer. Even a washing stool is there with a dishpan, and a beautiful stove with baking oven and water container and several hearth lids on each opening. And naturally there are all the right utensils and small implements in the cabinet. The cake platter on the table looks like our big one: blue plums on a stoneware plate!—In front of the window are flower boxes, and pictures hang in the corner, and a clock stands on the cabinet" (From a mother's letter to a friend, Munich, Christmas 1938).

Doll kitchen, Thirties.

Doll kitchen, Twenties. Housing
(height 43, width 93, depth 45 cm)
of lacquered wood, linoleum floor.
Two opening windows with
counterweights. Furniture
lacquered white, simple
construction. Stove: lacquered
white and nickel-plated, electric,
aluminum cooking pot. Utensils:
ceramic, porcelain lacquered
sheet metal. Special equipment:
closet with sheet metal racks.
(Historical Museum,
Frankfurt/Main)

"My Doll Kitchen", circa 1925.
Paper, cardboard, cutout game.
(Josef Scholz & Co., Mainz)
29.5 x 38.5 cm.

Most doll kitchen of the time were, like the one described here, set up as living room-kitchens with a few modern improvements. Customary, for example, were the working windows as mentioned, with small bolts, or the sash windows typical of the time, running water and/or a working sink with removable lower part holding two bowls, modern implements like wall-mounted coffee mills, crank-operated bread machines, mincing machines, pressure cookers or thermos bottles. The furniture, and often the floor too, were covered with linoleum.

Children with electric stove in the
"Children's Paradise" of the
National Garden Show, Essen,
1938.

Doll kitchen, late Thirties.
Housing (height 35, width 80
tapering to 67, depth 43 cm) of
colorfully papered wood, floor of
oilcloth. Furniture: lacquered
wood, tan; closet with built-in
porcelain racks. Water: sheet-metal
basin with faucet. Stove: black
lacquered or nickel-plated "gas
stove", for alcohol or "Teelicht";
nickel-plated cooking pot.
Utensils: aluminum, moulded
glass, clay.

Doll kitchen cabinet, late Thirties. Wood, lacquered in ivory and set with imitation natural wood, "etched" windows, built-in porcelain racks. Height 22.5 cm.

Doll kitchen, 1938. Amateur photograph.

The alcohol-burning "coal stove" was then replaced by the alcohol-burning "gas stove". The little working electric stove often still looked like a coal stove, but was already available in very modern form. The firm of Märklin Brothers, of Göppingen, put a 39 cm high, white enameled table stove with side baking ovens, on the market in the Twenties; a child could cook easily on it. In a time when functionality was highly valued, doll kitchens were desired that were suited to the size of the children, so that they could even get part of the way into them.

The Kitchen of World War II

The productive activity of the woman in the household was reckoned into the economic planning of the Nazi regime: knitting and sewing the family's clothing, making the household linens, growing fruit and vegetables, harvesting and preserving and finally keeping small animals all seemed to be a way of increasing the "people''s wealth''.

A basic education in homemaking and theoretical study of the kitchen and its implements were thus promoted by the state.

The war conjured up by the regime demanded of women, in addition to the role of the perfect housekeeper, a return to an occupation, for the economy could not do without their part of the work force.

The kitchen of the time was simple and without frills, and certain labor-saving measures had been retained from the Twenties, but the clearly rational kitchen was no longer postulated. The living room-kitchen with individual pieces of furniture, a big table, and perhaps a corner bench was much preferred. So was strongly veined natural wood, or ivory and lime-green lacquer, set off in part by intense red, but particularly farm-style painting. The "Advice for the Dowry" given in the 1938 book, "A Girl Wants to Marry'', are full of information: "A thousand Marks have become the basic concept, as seen in the marriage loan, of what a young couple needs for the basis of its household. With the help of the marriage loan (about 600 Marks, author's note), this sum will usually be attainable. A nice cozy living room-kitchen and a well-made bedroom can be had for that much. The living room-kitchen has the advantage of meaning a saving on fuel in the winter. Whoever doesn't want to live in the kitchen should set up a simple kitchen and a bed-sitting room.[22]

The simplest furnishing of the living room-kitchen for 1000 Marks looked like this: "Living room-kitchen consisting of kitchen cabinet, table and sideboard, couch with cushions and two chairs for 290 Marks."

A total of 165 Marks was suggested for kitchen utensils and dishes. The included machines were a food chopper with grater attachment, a coffee mill and a can opener.

The best apartment furnishings, for 3,500 Marks, included a "living room-kitchen in red Japanese lacquer with ivory lacquer". Belonging to it were: "1 kitchen cabinet, 1 combined dishwashing and work table with removable work surface, 1 utensil closet, three chairs with linoleum seats, 1 corner bench, 1 dinner table" for a total of 392 Marks; 545 Marks were to be spent on kitchen utensils and implements; the sum included, among others, a bread slicer, of course a food chopper with accessories, a kitchen scale etc., a water heater, a vacuum cleaner, an icebox and a (hand-powered) washing machine.

Such "a shining new arrangement, completely modern, is the dream of most young brides. But though one must save, still this dream will not always come true... Unmodern furniture can often be made to look like new without spending very much money, if one removes senseless decorations, opulent ornaments and overdone mouldings ... It is easiest to make hall and kitchen furniture out of old components. Simple smooth cabinets can be repainted, tables made higher or lower, accoding to the size of the young wife who is to work at them ... One can take a piece out of the middle of an unmodern kitchen cabinet, so that the upper part sets smoothly on the lower part and the cabinet takes on a practical height. Or one can also make good use of the space in between through the use of built-in glass or metal racks. Thus one can help oneself in many ways with the solution 'new out of old'."[23]

The age groups who married during the war really had to "help themselves", for even furniture was rationed in those days. After the "final victory", of course, the individual would have no more choice, not even as to his kitchen: according to a decree of the Führer on November 15, 1940, 300,000 apartments, mostly for young married couples, were to be built in

"Souvenir of the hundred-gram roast, October 1941". (Kitchen furnishings of 1934) Amateur photograph.

the first year of construction after the war. Their furnishing had already been planned in advance. In "The Beauty of Living", published by the German Work Front in 1941, the introduction states:

"...the problem of home furnishing (is) grasped by the dynamic of the National Socialist outlook on life. No longer are fashions or accidents of 'personal taste' the decisive factors therein, but the fact that the furnishings of a dwelling, the household goods, are the cultural possession of the people in the highest sense, so that all our race's requirements and cultural needs of the new times have to be taken account of . . .

"The moment will come when nothing but 'German housewares' will be available, and all sidetracks of fashion must cease to appear because of the waste of work force and material they cause!"[24]

Kitchen, circa 1940. Amateur photo. (Photo Archives of E. Maas, Frankfurt/Main)

Doll Kitchens of World War II

For children of wartime Germany—as in the big kitchen—old items, when available, were brought out and modernized, or leftover scraps of wood were used to make things, for there was less and less to buy. In the end, making it oneself became the only possible way of giving children anything, once the government banned the use of glass, ceramics and wood for toys in March of 1943.[25]

There was no lack of do-it-yourself directions in women's and family publications. For example, one fashion magazine made the following suggestion in its 1940 Christmas issue: ". . . when everything is sawed out, glued together and well dried, we proceed to the painting of the individual objects: cabinet, table, bench and chairs look particularly nice when they are painted cobalt or cornflower blue. We'll also set off the appropriate decorations in glowing colors: wreaths and borders. We'll keep the hearth brick red, with white lines to mark the individual bricks." The peasant kitchen described here had a hearth for an open fire, over which big brass kettles hung in the flue, and a water bench with two pitchers. The idealized and romanticized peasant milieu was typical of the times.

Along with these painted kitchens, there were simple rough wooden kitchens with an increasing lack of color, but with a few flowers in dull colors (glued on).

Different are the wartime kitchens with ceramic pots finished in cream, red-brown or lime green, that were sold in the government's welfare "Street Collection" in Magdeburg-Anhalt in January, 1940, and in South-Hannover in January, 1942. (illustration, page 130).

Peasant doll kitchen, circa 1940.
Made from directions in a
handicraft magazine.

Doll cooking utensils, circa 1900.
Enameled iron, spotted pattern.
Height of the pot 9 cm.

Built-in kitchen made by the
Poggenpohl firm in the Fifties.
(Photo: Poggenpohl, Herford)

The Modern Kitchen

The very first kitchens of postwar times were makeshift kitchens with all kinds of utensils that one could make oneself out of tinplate, or put together out of leftover stocks of military surplus items. Since the money was worthless, business people held back existing stocks. Only after the 1948 currency reform did old kitchen buffets and stoves of the Thirties and Forties pop up in stores. They consisted not only of hoarded items, but at first were produced anew in old shapes and styles.

Furniture storeroom with kitchen furniture, 1949. Photo by Kamolz.

Doll kitchen, circa 1905. Housing (height 42.5, width 89.5 depth 43.5 cm) of wood, papered in tile pattern. Furniture: lacquered wood, built-in cabinet and shelves. Water: water bench with pitcher and bucket. Stove: stamped sheet iron, doors and lion's feet of brass, for spirit; tinplate cooking pots, additional "gas stove", cast iron, lacquered, for alcohol. Utensils: light blue enameled sheet iron, porcelain, earthenware. Special equipment: broom closet.

Doll kitchen, 1905-10. Housing (height 37, width 73, depth 43 cm) of wood painted in tile pattern. Furniture: cabinet and sideboard painted light beige, bronzed. Water: sink, bucket. Stove: sheet iron with stamped brick pattern, gold doors and legs, sheet metal cooking pots. Utensils: tinplate, gold-bronzed canisters, painted porcelain. Special equipment: sheet-metal broom rack.

Kitchen stoves and ovens, 1949.
Photo by Kamolz.

With the revived economy of the early Fifties, the "modern" type of kitchen developed. It was influenced on the one hand by America, and on the other very clearly by the kitchens of the reform years between 1925 an the early Thirties, for after the second World War, as after the first, there was a great housing shortage. The small kitchen type of the Twenties, set up to be strictly functional, also met the needs of the Fifties. The requirements of that time, practicality in the arrangement of the furniture, practical built-in features, mechanization and electrification of kitchen appliances, could now become reality for a wider range of the population, more slowly, to be sure, than one might wish. The mail-order house of Neckermann in

Frankfurt/Main still offered in 1957-58 various old-fashioned natural-stained kitchen sets with buffet, table, trunk-bench and chairs, cleaning closet, towel rack and dishwashing table, covered with "brown/yellow granite" linoleum. But the first "Swedish style" kitchen buffets in pastel color combinations, "dove gray/soft pink/ivory" with sliding doors, fully enclosed base and plastic-covered counters were offered, as well as a "modern built-in kitchen—to save time and strength". In 1960-61 the brown wood furniture gave way to the built-in kitchen with plastic front "in the living style of our time", with beveled hanging cabinets, a moulding with supply racks underneath, and imperceptibly integrated large appliances such as refrigerator and stove. Three years later there was only a single brown buffet in the catalog. The "New Kitchen" with handleless fronts in pure white, already produced by all the leading kitchen manufacturers in the Mid-Fifties, had now, after a slight delay, established itself at the mail-order house "for everybody". The originally very expensive appliances, mixers, kitchen machines and washers, began to be taken for granted.

The increasingly perfected and systematized kitchen took on the atmosphere of a laboratory. But in the Sixties the severity of technology began to be softened again with colors; later imitation wood came to be used, and finally "rustic kitchens" with pleasantly curved profiles of real wood. And today things emerge again that were stuck behind doors sixty years ago to protect them from dust and spattering fat, they hang on the walls and stand on the cabinets. "The Kitchen from Grandmother's Time", with its old and copied implements promises lost warmth and familiarity. Of course, there is no more shortage of electric kitchen machines and modern major appliances. "Absolutely unseen", hidden behind old-fashioned fronts, are refrigerators and freezers, dishwashers, stoves, ovens, grilles and all kinds of highly modern kitchen appliances.

The need to "flock together" while cooking and eating, experienced by young people in communes, has given the living room-kitchen a future again.

Above: doll kitchen with larder, circa 1910. Housing (height 38, width 102, depth 37.5 cm) of wood lacquered in tile pattern. (Older housing, circa 1870, with dining room added later). Furniture: lacquered wood. Water: reservoir with brass faucet and sink. Stove circa 1920: white-enameled sheet metal, electric; light blue-enemeled cooking pots. Utensils: light blue enamel, ceramic pots, porcelain. Special equipment: lid and spoon holders, wall-mounted coffee mill; in the larder: egg cabinet, ladder chair, petroleum-measuring apparatus.

Below: doll kitchen, circa 1910. Housing (height 34, width 67, depth 38 cm) of lacquered wood, floor papered in tile pattern. Furniture: lacquered wood, patterned. Water: reservoir with faucet, sink. Stove: "gas stove", lacquered sheet metal for candle, sheet-metal water kettle, lacquered. Utensils: blue and white speckled enamel, porcelain.

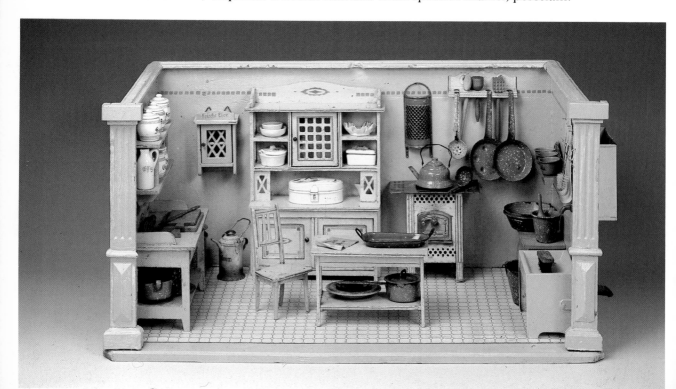

Above: doll kitchen, 1910-20. Housing (height 55, width 122 tapering to 94, depth 53 cm) of wood lacquered with tile pattern. Furniture: lacquered wood, palace-like cabinet. Stove (shown above): sheet iron stamped with tile pattern, bronzed, brass doors, enamel cooking pots speckled in gray and white, for alcohol. Utensils: canisters of porcelain with gold, porcelain coffee service with adhesive pictures, enamel utensils speckled gray and white. Special equipment: champagne cooler (at right).

Below: doll kitchen, 1910-20. Housing (height 39, width 75, depth 41 cm): wood with tile papering. Furniture: lacquered wood, kitchen cabinet with glass windows. Water: faucet with bucket, no water connection. Stove (at right): stamped sheet iron, bronzed, doors and legs brass, for alcohol; aluminum cooking pots. Utensils: aluminum, ceramic, printed sheet metal. Special equipment: broom rack (right rear).

The Modern Doll Kitchen

In the first postwar years people were happy with small, modest doll stoves and furniture of tinplate or aluminum, painted in black, dark gray, tan and dull red hues. Capable people had set up small workshops in the do-it-yourself manner, using materials taken from wartime leftovers. The style of these small pieces was still that of the Thirties and Forties. When industrial production also began again, old models, for which models, forms and tools were available, were built at first.

Doll kitchen, Fifties, reminiscent of the Twenties. Typical of the Fifties: cutoff hanging cabinets with racks below, electric stove with straight bottom, continuous counter with integrated sink. Housing: wood, folding together, papered with black and white tile pattern. Furniture lacquered ivory. Privately owned. (Photo: Gerd Schroth, Frankfurt/Main)

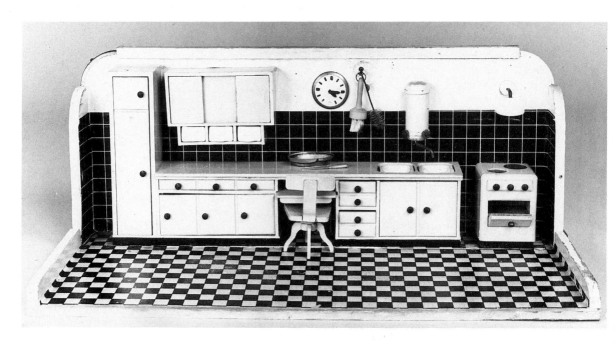

Doll kitchen, early Fifties. Housing (part of a double room, height 21, width 34.5, depth 30 cm): wood, papered in lime green and bright yellow. Furniture: natural wood, celluloid window with anodized metal parts. Imitation "Niroxa" sink. Utensils: plastic.

Doll cooking utensils, circa 1948. Cast aluminum; height of the pot at rear, 3.5 cm.

Electro-stove, Fifties. Aluminum, painted beige and black, handmade. Height 10.2 cm.

At first there were scarcely any complete kitchens. Only in the early Fifties were small wooden kitchens produced. At the same time, the first "modern" sheet-metal stoves, furniture and appliances appeared. Most of them were finished in ivory, some printed in several colors. For example, on the oven doors there were color pictured of roast chickens or baked goods, or on the wall behind the sink a printed window with landscape and kitchen curtains. Modern appliances such as washing machines, clothes dryers and grilles were also produced for doll kitchens now. The combination of all these individual parts, the "Elctro-stove" (running on wood alcohol), the sink with hot water heater and the hanging and under-counter cabinets for the "kitchen row" were a logical development already known in the big kitchen. This space-saving flat kitchen has become the accepted modern form of the doll kitchen. There are naturally exceptions such as the

Doll kitchen "in a closet", late Twenties. Housing (closed: height 39, width 50, depth 11) of wood with tile papering. Furniture: built-in except for table and stools. Water: boiler, sink. "Gas stove", lacquered sheet metal, for alcohol; enamel cooking pots. Utensils: printed sheet metal, porcelain. Special equipment: fruit bowl, stoneware platters with nickeled rims.

Above: doll kitchen with bathroom, Twenties. Housing (height 30, width 69, depth 34.5 cm) of wood, walls with tile-patterned oilcloth, linoleum kitchen floor. Furniture: lacquered wood. Water: faucet (no reservoir), porcelain double sink. In the bathroom: hot water heater for "gas" of copper. Stove: wooden block, details painted to look like brass; bronzed sheet metal pot. Utensils: porcelain.

Below: Doll kitchen, Twenties. Housing (height 39, width 87, depth 45 cm): tiled walls behind the stove and sink, opening window with original shade. Furniture: lacquered wood, table and cabinet counter with celluloid tops. Stove: lacquered sheet metal; enamel cooking pots. Utensils: porcelain, ceramic. Special equipment: washing machine.

"Kitchen row" for a doll kitchen
of the Fifties; wood, with sliding
doors, lacquered bright yellow and
soft pink. Height 14 cm.

tiny kitchens of the small doll houses of those days, and
the "Lego Kitchen" of the late Sixties. The function of
the kitchen, cooking, plays no role here. In the modern
building-block system the cabinet pieces can be
assembled as one wishes, complete with opening doors,
shelves, stove and sink.

Doll kitchen row, circa 1970. Back
wall with exhaust fan and hot
water heater. Sheet metal, printed
in bright colors with shades of
orange predominating. "Electro-
stove" for dry alcohol. Utensils:
aluminum cooking pot and ladle
set. Set of bowls and cooking pot
printed on the back wall, as are
colorful utensils, roast goose etc.
in the lower cabinets. Made by the
Schopper toy firm for the
Bauknecht kitchen firm. Height 28
cm.

Color plate
Above: doll kitchen, Twenties.
Housing (height 42, width 85
tapering to 65, depth 36 cm) of
painted wood, tile-patterned
papered floor; opening window.
Furniture: lacquered wood. Stove:
lacquered sheet metal, for
"Teelicht" (solid alcohol fuel);
aluminum cooking pots. Utensils:
porcelain, glass. Special
equipment: wall shelf with
porcelain racks.

Below: Doll kitchen, circa 1930.
Housing (height 36, width 73,
depth 36 cm) of lacquered wood,
Stragula (?) floor, two windows
near the corners. Furniture:
natural wood, lacquered,
removable sink. Stove: lacquered
sheet metal, handmade. Utensils:
aluminum, enamel, porcelain,
pressed glass.

Doll kitchen sink, circa 1970.
Sheet metal, rear wall colorfully
printed with imitation window
and tiles. Height 29.5 cm. (Plastic
hot water heater and soap dish
made by Crailsheimer). Firm of
Martin Fuchs, Zirndorf near
Nuremberg.

Part of a doll coffee service, circa
1960. Porcelain with adhesive
pictures. Height of the coffeepot
11 cm.

Upper left: stove, Fifties. Sheet metal, ivory with blue lines, oven door printed with baked goods. Sheet metal and aluminum utensils, height (with back wall) 21 cm.

Upper middle: doll refrigerator, Fifties. Stamped sheet metal, celluloid sliding doors. Height 20 cm.

Upper right: "Electro-stove", Fifties. Stamped sheet metal, white and black with red metal and plastic parts, clock, aluminum utensils; for dry alcohol. Height (with back wall) 18.5 cm.

Below: rotary grille, makes sparks, Fifties; sheet metal, partly lacquered red, wheels colorfully printed, roast and chicken of paste. Height 18 cm. (Firm of Martin Fuchs, Zirndorf near Nuremberg)

Cooking has moved completely out of the doll kitchen into mother's big kitchen. Many children are allowed to cook candy or other good things on the big stove at an early age, and sometimes even to bake cakes for the whole family.

In addition, there are relatively large kitchen implements with which the child can begin. In 1977 a large German mail-order house offered a "children's kitchen cabinet with sink, really working water faucet and shelf", almost one meter high, a two-burner electro-stove "for cooking and boiling" and an electro-stove with two burners and baking oven "working like a real one! Can be turned on separately, 220 volts, 150 watts, VDE tested. Safety thermostat, transparent window, Schuko wiring, non-rusting burners, approx. 18 x 21 x 23 cm in size." One can also buy battery-driven hand mixers, working automatic coffeepots, and as before, baking sets as "a genuine playing and learning asortment".

Lego doll kitchen, circa 1970. Height of the kitchen cabinet 8 cm.

The old idea of getting children used to their later roles in life through play has survived, but its center of gravity has moved with the times. If one walks through the toy departments of department stores, it becomes clear how few such "housewifely" toys are offered in comparison to the vastness of other toys. The wishes of most girls in the realm of dolls are more likely to center on a "beauty salon, in which one can color hair" or a "jewelry boutique for Barbie".

The independent working woman who also keeps house with the help of appliances is the model, not the housewife. So doll kitchens can have no relevance today. New trends may be seen in the "Nostalgic Doll Kitchens" that imitate Grandma's kitchen; but that means it is Grandma's kitchen that is copied here.

Doll kitchen model, 1982. Alfa-System Kitchens, Wiesbaden. (Photo: Gerd Schroth, Frankfurt/Main)

The Pantry

The cold-cellar was formerly the place to keep all heat-sensitive food supplies. Here, along with the potatoes stored for the winter, the root vegetables in sand and the apples and winter pears, homemade things were also stored in the wooden drawers: sauerkraut, pickles, cranberries and prunes in gray-blue stoneware crocks, eggs (in water glasses) and lard in brown Bunzlau containers. But fresh foods were also put in the cellar, covered with cheesecloth, fly covers or pieces of cloth, for only a few households had a pantry.

Only in the Seventies of the Nineteenth Century, when large apartment houses were built in the cities and the small cellars of the individual tenants were only big enough for wood and coal, was it all but impossible to get along without a pantry. In the floor plans of the larger apartments, the pantry was roomy and usually situated next to the kitchen at the north; in small apartments it was just a tiny closet.

From this time on, doll kitchens also had pantries built onto them more and more frequently. The furnishing was simple: a cabinet with drawers, shelves, perhaps a table, a ladder chair, egg containers, storage jars, stoneware and porcelain bowls, canisters of all kinds, even for vinegar, oil and petroleum.

In Thuringia the big folding wooden rack for cooling flat round cakes was also kept here.

Usually there was a food locker with gauze or at least a fly screen, and almost always small linen sacks in which dried tea, mushrooms or fruit could be hung in the air. In the "better" pantries, whether big or small, was the icebox. Its inner capacity was scant despite its outer size, for much space was lost to the hanging zinc box with blocks of natural ice and the thick walls needed for insulation.

Doll kitchen with pantry, 1899.
(Herz & Ehrlich, Breslau)

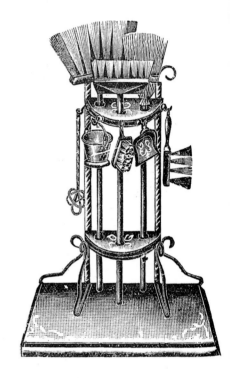

Broom rack, circa 1913. (Borho, Baden-Baden)

Even in doll kitchens, these iceboxes worked with a small piece of ice.

During the course of the Nineteenth Century, food storage gradually moved from the private household to the dealer. In the city household of today, where keeping potatoes in the cellar, common until after World War II, is no longer done, and where generally only the weekly or even daily needs are bought, this development has reached its high point for the time being. Small modern apartments naturally have no pantries any more—they had already disappeared from doll kitchens in the Twenties—and the refigerator and freezer have replaced them.

These refrigerators, which took the place of the old iceboxes, and which could produce their own coldness with the help of a cooling agent and energy (coal!, gas, electricity), had already been invented by the end of the Nineteenth Century. In American households they had already been introduced in 1916-17, in Germany in the Twenties.[26] Modern refrigerators for doll kitchens were first produced in the Fifties. They were perfect in form, had a defrosting compartment, shelves, door racks and even interior lighting, but unlike the old iceboxes, they no longer worked.

"Rex pressure cooker" for
children, circa 1910. (right) Height
of the pot 18 cm. (Historical
Museum, Frankfurt/Main)

An important invention for keeping household foods, which came into its own in children's rooms as well, was the pressure cooker for home canning. The "Wecktopf" canner was invented around 1895 by Johann Weck of Olfingen, Baden, but was based on older developments.

The Weck canner was such a great success that as early as 1910-20 the pot and the equipment that went with it, the glasses, seals and thermometer, were imitated by other firms and quickly put on the market as children's canning apparatus.

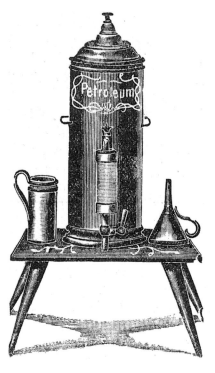

Petroleum dispenser, circa 1913. (Catalog of H. Borho, Baden-Baden)

Kitchen cabinet, completely equipped, circa 1913. (Borho, Baden-Baden)

Who Made Doll Kitchens?

Advertisement of the Nuremberg metal goods factory, Bing Brothers, in a housewives' magazine, circa 1890.

Most manufacturers of doll kitchens are unknown. Doll kitchens are not marked, and even with catalogs they are hard to identify. The sheet metal kitchens of the renowned tin toy manufacturers are an exception, since their catalogs have been widely reproduced. But a look at the offerings of such a firm, as for example the Märklin Brothers of Göppingen, can make clear the difficulty of researching. The Märklin Brothers sold not only sheet metal kitchens but also wooden ones. The latter could hardly have been made in their own factory. Were they taken over from the production of a woodworking toymaker? Or were they specially made to the specifications of the Märklin Brothers? Which firms actually made them?

The big question mark remains. Like Märklin, other firms also included articles from outside in their sales programs. This was done to the extent that not only such articles as the firm itself did not produce were taken over, but so were those that others made more cheaply than the firm reckoned it could. That means that some manufacturers were simultaneously vendors of the goods of other firms. If even the manufacturers' own catalogs are not always completely clear, one can imagine that it is even harder to tell the manufacturers from a dealer's catalog.

The catalogs that point to the manufacturers are those that include only a firm's own products or can be relied on because of what is known of the firm's history, plus advertisements that show a firm's own goods. Other sources are works from the late Nineteenth and early Twentieth Centuries on the subject of economic science, which dealt thoroughly with the home products industry in general and the toy industry in particular.

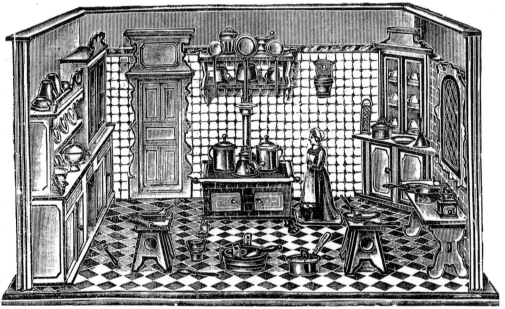

Upper: wooden kitchen, 1895.
(Märklin 1, page 102)

Lower: doll kitchen, circa 1913
(Borho, Baden-Baden)

Manufacturers of Wooden Kitchens

"Dolly's Kitchen", do-it-yourself instructions in a housewives' magazine of 1923.

Wooden kitchen, 1895. (Märklin 1, page 102)

Doll kitchen, 1899. (Herz & Ehrlich, Breslau)

In addition to series production by the toy industry, parents and children have always made doll kitchens of wood themselves. There were many directions for such work published in housewives' and family magazines and activity books for children.

Doll kitchens in great quantities came from the Nuremberg area and the Erzgebirge. Various doll kitchens of wood are pictured in Hieronimus Bestelmeier's work (Nuremberg, 1803) and the Nuremberg pattern books of the Mid-Nineteenth Century, while Jacob Seyfried, a manufacturer from Fürth, near Nuremberg, displayed an entire "Furniture Assortment of Wood" at the 1844 German Trade Fair in Berlin.[27]

At that time people often had doll kitchens made by artisans. There seem to have been small factories that grew out of carpentry shops in many cities, especially in South Germany. One example that can be named is August Götzinger of Merseburg, who in 1843 offered "from our own factory kitchens without utensils, but otherwise with all furnishings, 14-18 inches, 15 to 20 Groschen apiece"; as well as kitchens of the same kind, "20-28 inches wide with front peak, columns and decorations of paste attached, also with glass cabinets in the larger ones, also with feet, 1 Thaler to 1 Thaler 20 Groschen apiece".[28] In Nuremberg, for example, there is proof that Carl Bierhals produced doll kitchens in 1906.

The center of wooden doll kitchen makers has been for many years the area around Grünhainichen, particularly Eppendorf in the Erzgebirge. In a guide to the Saxon-Thuringian export industry, published in Dresden in 1897, three Eppendorf firms specifically advertised doll kitchens or furniture: Richter & Wittich,

Paul Leonhardt, and Lehnert & Co. Around 1911 there were fourteen firms in this city that made chiefly doll kitchen housings and furniture.[29]

Artistically impressive toys that sought to to influence the taste of children in the framework of the reform movement and the changed esthetic consciousness of the times around 1900 were produced in many different workshops. Among them was the Wooden Toy Factory in Grosszolbersdorf, Saxony, which produced so-called "Dresden Toys" to artists' designs, including in 1910 a very modern and cheery little kitchen with smooth cabinets and modest furnishings,[30]; another was the Hessian Toy Manufacturer, founded by the architect Conrad Sutter, which displayed doll kitchens among other products at the 1914 Easter Fair on Leipzig.[31]

In the first years after World War II, the toy industry at first produced their old models from prewar days again. Shortly after the currency reform (1948) the first German Toy Fair was held in Nuremberg, at which firms such as Herbert Leonhardt KG of Wallenfels, Upper Franconia, and Fritz Altmann of Bad Driburg, Westphalia, offered kitchens, almost all of them made of wood and finished in lacquer. Utensils were produced not only of aluminum and sheet metal, but also of plastic.

Doll kitchens in various sizes,
Mid-19th Century. (Nuremberg
Toy Catalog)

Manufacturers of Sheet Metal Kitchens

Around the middle of the Nineteenth century and later, the firm of Rock & Graner in Biberach on the Riss produced sheet metal kitchens. Gottfried Striebel also worked for this firm, in whose nicely colored pattern book from the same time appears a sheet metal kitchen with a painted brick-type stove under a large flue, complete with chicken pens and all kinds of utensils.[32] In the Fifties and Sixties beautiful pattern books appeared in Nuremberg, including sheet metal kitchens.

Sheet metal kitchen and stove, 1850-60. (Nuremberg Pattern Books, page 41)

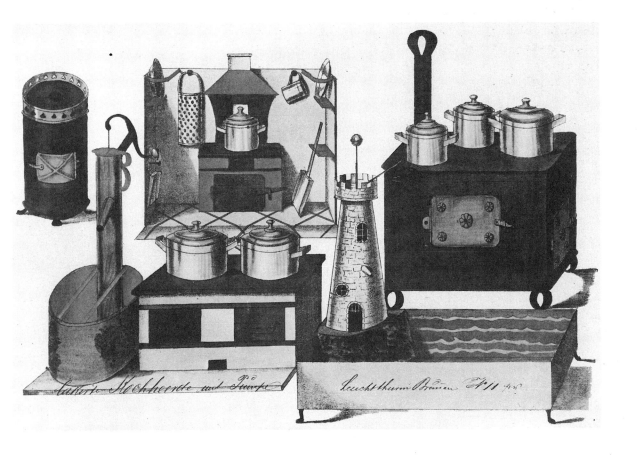

Sheet metal kitchen, 1895.
(Märklin 1, page 223)

In Württemberg and the Nuremberg-Fürth area were the most, and the most important, producers of sheet metal toys, and thereby also of sheet metal doll kitchens, in the Nineteenth Century. At the Vienna World's Fair in 1873, only two firms other than Rock & Graner of Biberach and C. Martins of Berlin displayed goods of sheet metal, including doll kitchens or utensils, and they were C. Henglein and Ph. Wüstendörfer, both of Fürth. And the Nuremberg metal goods factory of the Bing Brothers should not be forgotten.

In addition, the Göppingen firm of Märklin, which was to become famous because of their toy trains, produced toys for the doll kitchen when they began in 1859, and when the Märklin brothers took over the firm in 1888, they at first made furnishings for children's kitchens and cookstoves. In their main catalog of 1895—in which trains and sheet metal cars already played a large role—Märklin offered "kitchens, white lacquered sheet metal, with furnishings and running water" and "folding kitchens with running water, without furnishings".

The sheet metal toy factory of Clemens Kreher, in Marienburg, Saxony, is represented in the guide to the Saxon-Thuringian export industry of 1897 by "children's cookstoves, dull-angled household utensils" and "kitchens".

Sheet metal kitchen, 1895 (Märklin 1, page 223)

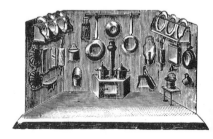

Doll kitchen (Stukenbrok, Einbeck, 1912, page 131)

"Kitchen furnishings, white lacquered sheet metal" (Ullmann & Engelmann, Nuremberg, circa 1900, page 2597)

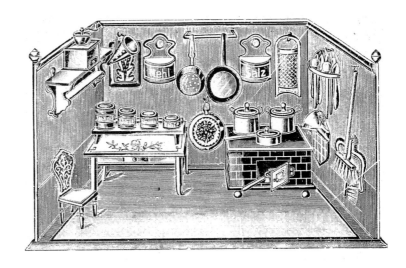

Sheet metal doll kitchen, finely lacquered, furnished with wooden furniture (Moko, Nuremberg 1928-30, page 2164).

Sheet metal kitchen, lacquered, fully furnished (Moko, Nuremberg, 1928-30, page 2164)

The firm of Ullmann & Engelmann of Fürth made sheet metal kitchen furnishings in 1901, and Moses Kohnstam of Nuremberg still offered a flat sheet metal kitchen and a larger one of the Märklin type in 1928-30.

Doll kitchen, circa 1900. Housing added (height 30, width 52, depth 38 cm) of colorfully lacquered wood, made singly. Two etched windows. Furniture: sheet metal, lithographed in bright wood finish, with kitchen sayings in black. Cabinet with fly screen. Stove: sheet iron, brass-colored door and lion's feet, tinplate cooking pots with brass lids.

Color plate
Above: doll kitchen, Thirties. Housing (height 49, width 109, depth 54 cm) foldable: wood, papered wall, linoleum floor, opening window. Furniture: lacquered wood set with imitation natural wood, etched windows. Water: flat reservoir set in the right wall, basin with water faucet. Sink with turning "drawer". "Gas stove": sheet metal, for alcohol, on lacquered sheet metal table, aluminum cooking pots etc. "Kitchen wonder" and pressure cooker. Utensils: porcelain canisters and coffee service, Bunzlau pots. Special equipment: food grinder, bread slicer, thermos bottle.

Below: doll kitchen, Forties. Housing (height 26, width 75.5, depth 37 cm) of lacquered wood. Two windows. Furniture painted in peasant style. Stove: sheet metal, doors painted on, handmade; cooking pots of sheet metal. Utensils: WHW ceramic.

    After World War II, Kindler & Briel (Kibri) of Böblingen made doll stoves as well as sinks. The greatest distribution of the time was probably that of the Martin Fuchs firm of Zirndorf (MFZ), near Nuremberg, with their sheet metal kitchen furnishings of all kinds.

The Trade

The sale of doll kitchens, like that of other toys, took place via the distributors, located mainly in Nuremberg, but also through the yearly fairs and the great exhibitions of the Nineteenth Century—from toy fairs to world's fairs.

The final consumers, the parents and children, saw and bought the toys from toy dealers and found alluring offerings—especially in the time before Christmas—in advertisements in the family magazines as well as the catalogs of mail-order houses. The main German firms were Mey & Edlich of Plagwitz, Lepizig, and the Versandhaus August Stukenbrok in Einbeck.

Cookstove packaging "with great saving of space" (Moko, Nuremberg, 1928-30, page 2165)

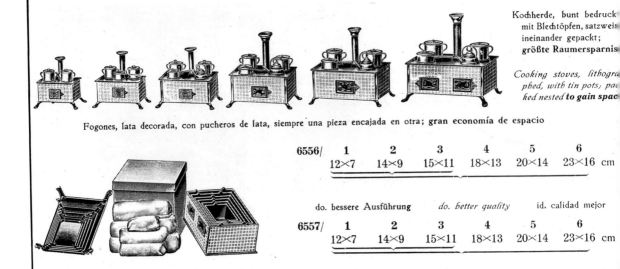

Kochherde, bunt bedruck mit Blechtöpfen, satzweis ineinander gepackt; größte Raumersparnis

Cooking stoves, lithogra phed, with tin pots; pac ked nested **to gain spac**

Fogones, lata decorada, con pucheros de lata, siempre una pieza encajada en otra; gran economía de espacio

6556/	1	2	3	4	5	6	
	12×7	14×9	15×11	18×13	20×14	23×16	cm

do. bessere Ausführung   *do. better quality*   id. calidad mejor

6557/	1	2	3	4	5	6	
	12×7	14×9	15×11	18×13	20×14	23×16	cm

Crate with an assortment of toys
(Printemps, Paris, 1891-92)

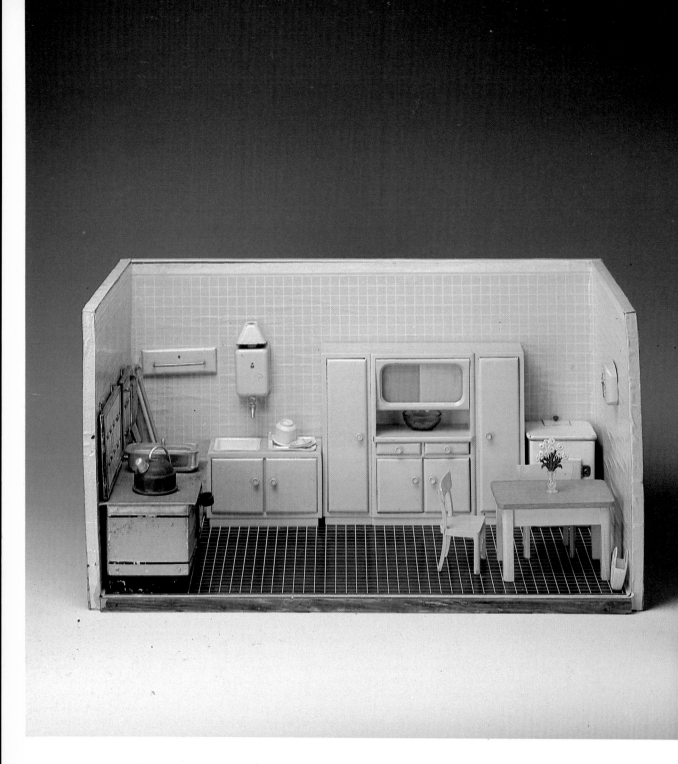

Doll kitchen, circa 1950, in the "Reform Kitchen" style of the Thirties. Housing (height 32.5, width 64, depth 34.5 cm): wood with tile-patterned wallpaper. Furniture lacquered, two-colored glass in window sashes. Water: boiler with double sink and cabinet under it. Stove: sheet metal, lacquered, handmade—electric. Special equipment: refrigerator of porcelain with wood top.

Above: doll kitchen, early Thirties, with two sets of furnishings. Housing (height 44, width 118.5 tapering to 102, depth 42.5 cm): lacquered wood with linoleum floor, two opening windows. Furniture: natural wood covered with linoleum. Stove: electric, lacquered and nickel-plated sheet metal, nickel-plated cooking pots. Utensils: porcelain. Special equipment: embroidered tablecloth, hand towel with saying.
Below: Sixties furnishings. Housing (see above) was redone for the daughter. Furniture: lacquered wood covered with plastic. Stove: sheet metal, oven door printed, for dry Esbit; aluminum cooking pots. Utensils: porcelain with adhesive pictures. Special equipment: clother dryer, lacquered sheet metal with copper drum, washing machine with hand crank, refrigerator.

Doll Kitchens From Outside Germany

The individual doll kitchens, not integrated into doll houses, that concern us here seem to have been produced mainly in South German toymaking areas. The emphasis is on the word 'mainly', as there was production in other lands, at least in small volume for their own needs.

The French doll kitchens even had an importance that spread well beyond their own region. Many of them were exported. An 1891 toy price list of the famous Parisian department store, Magasins du Printemps, for the Scandinavian countries. It includes beautiful utensils packed in cases and boxes, as well as doll stoves and two small doll kitchens.

Wooden kitchen. (La Samaritaine, Paris, 1909)

Wooden kitchen. (La Samaritaine, Paris, 1914)

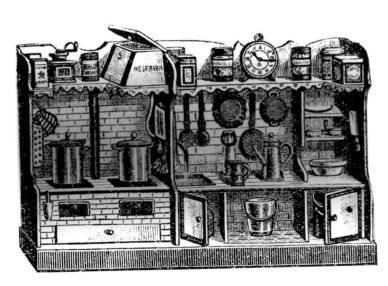

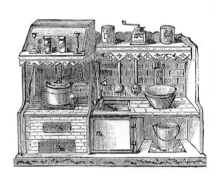

Wooden kitchen. (La ville de Saint-Denis, Paris, 1914)

English-type doll kitchen in a toy
store. Illustration in a children's
book, "The Wonder of a
Toyshop", circa 1835.

The most prominent characteristic of these French
kitchens, including the wooden ones, is their small size.
They are as flat as a closet, and their housings served as
packing boxes when they were shipped. Sometimes the
utensils were sewn onto the wooden kitchens through

Pantry, circa 1858, from the estate
of "Lullu" Röder Lorey,
Frankfurt/Main. Housing (height
42, width 78, depth 51 cm): wood,
originally red-brown, painted
ivory and light blue circa 1910.
Special features: blown jelly glases
and bottles, porcelain storage pots,
linen sacks with embroidered
lettering. Food carrier and sugar
breaker at left, stoneware bowls.
(Historical Museum,
Frankfurt/Main)

Furnishings of the pantry, circa
1900. From left to right: petroleum
dispenser, lacquered sheet metal,
cake-cooling rack (Sig. Blömer,
Frankfurt/Main), fly-free cabinet
and food cover, icebox with
bottles, lacquered sheet metal (Sig.
Blömer, Frankfurt/Main), egg
rack (Sig. Blömer,
Frankfurt/Main). Height of the
cake rack 29.5 cm.

holes bored in the wood, as was done with cardboard boxes. The furnishings were simple, sometimes primitive; the small stamped sheet metal utensils were painted with colorful alcohol lacquer.

The doll kitchens of the English-speaking world, The Netherlands and in part also Northern Germany, were not all sold by the Nuremberg toy trade. They often differ very noticeably from those produced in Southern Germany, especially by having a different type of stove. In these kitchens there is rarely a stove open at the sides and set under a flue attached to the back wall; more commonly it is the so-called pier-arch stove with side walls, and the flue is supported by the front of it, usually forming an arch. A stove of this kind, in a doll kitchen of the Schmarje family of Altona, is to be seen in the Altona Museum in Hamburg, just as such stoves are in Dutch or English doll houses. A book published in England in 1835, "The Wonders of a Toyshop", shows a doll kitchen with that type of stove. The Swiss doll kitchens differ from the South German too, but not so much in furniture and implements; it is rather the stove construction that differs here too. The flue in the middle of the back wall or in a corner is just like the German, but the stove itself has a particular form: on a typical masonry base with room below for the supply of wood—cut out in front and on the sides— a small hearth is built up with small open arches for making the fire. The flat surface on top has holes to hang the pots in.

Sheet metal kitchen with utensils. (Printemps, Paris, 1891-92)

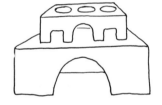

Sheet metal kitchen (Printemps, Paris, 1891-92)

What Can Be in a Doll Kitchen

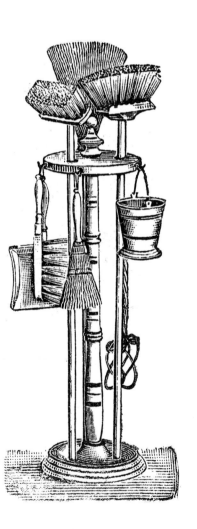

"Household goods" include everything that is needed in a kitchen and household; that means cooking pots and porcelain dishes, utensils and machines as well as cleaning things and the stove.

At first we are concerned with food and cooking utensils for dolls. They generally resembled the big ones and were usually made by the same firms. They were made of a wide variety of materials, for in the big household one could presumably cook many dishes only in very specific pots. This is affirmed by the rules printed in the introductions to many cookbooks:

"For rice and milk, use earthen vessels that cook "lightly" and do not scorch easily. Potatoes and fruits with skins should not be cooked in iron vessels, because they blacken in them and stay hard.

"Fish are best cooked in enameled pots, in any case in undamaged ones.

"Sauerbraten and sauerkraut must not be cooked in copper or brass vessels, nor in earthen pots with lead glazing, because acids can dissolve poisonous salts."

Earthenware, Fayence, pottery, stoneware, porcelain, glass, wood, copper, brass, tin, sheet metal, enamel and aluminum are discussed in the following chapters. More unusual metals too, that to be sure were not used often, have been found in the doll kitchen; for example, wrought and cast iron, nickel and zinc, as well as alloys such as Britannia and German silver, used mainly for silverware.

Broom rack, circa 1913. (Borho, Baden-Baden)

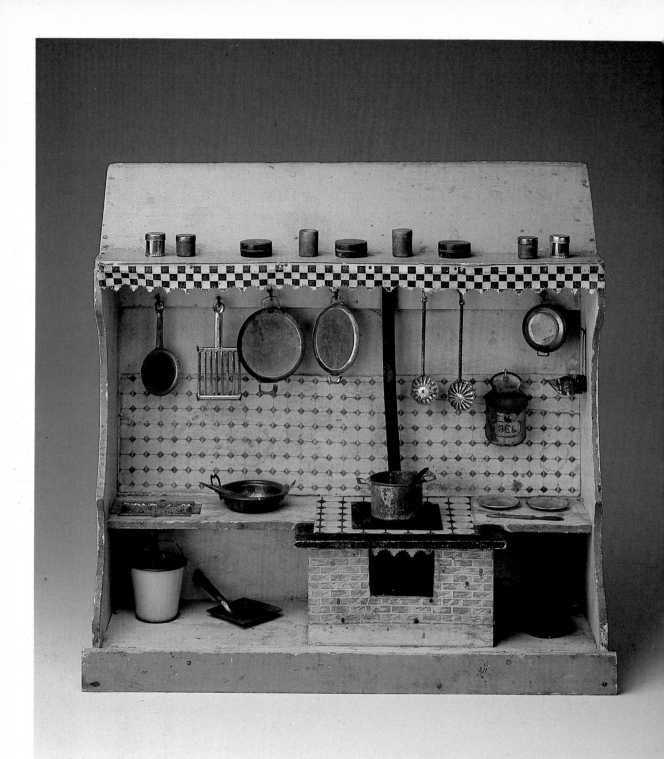

French doll kitchen, circa 1910.
Housing (height 37.5, width 40.5,
depth 13 cm): wood, papered in
tile and brick patterns, oilcloth
edges. Water: wash basin. Stove:
wood. Utensils: sheet metal
covered in oil paint.

Earthen utensils from between
1850 and 1920, mainly from Hesse.
Height of the wooden shelf 25 cm.

Still to be mentioned are the stainless steel utensils that were put on the market for big kitchens in the Twenties by WMF of Geisslingen. In the last decades they have been used less for doll kitchens than for bigger children's stoves in usable sizes.

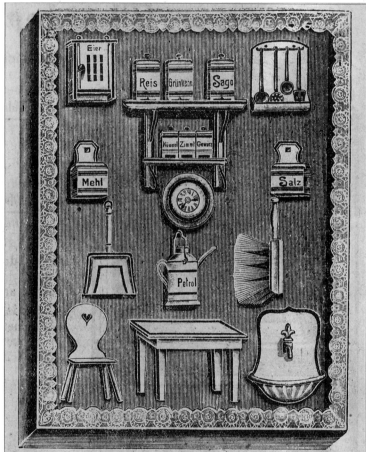

Kitchen set, circa 1913. (Borho, Baden-Baden)

Earthenware

Earthen three-legged pan, circa 1800, diameter 4.5 cm. (Historical Museum, Frankfurt/Main)

"Earthen utensils are the cheapest, but also the most breakable, and they can be dangerous on account of their glazes," (because of the lead content), it is said in a housekeeping lexicon of 1884.[33] On the other hand, many cooks swear by earthen cooking pots to this day for roasting meat. One need only think of the Roman pot. In the Eighteenth Century and the first half of the Nineteenth, earthenware of all kinds, roasting pans (known as "sow snouts") and tripod pans, cooking pots, milk jugs, plates and bowls were common even in well-to-do households. This becomes clear in the following famous story told by Goethe in "Dichtung und Wahrheit":

"There had just been a pot market, and not only the kitchen had been equipped for some time to come with such goods, but similar utensils had been bought in small sizes for us children to play with. On a beautiful afternoon when all was quiet in the house, I was occupying myself with my bowls and pots on the porch, and since I had run out of things to do with them, I threw one piece into the street and was happy when it broke so merrily. The Ochsensteins, who saw how I was entertaining myself and happily clapping my hands, called: 'More!' I did not hesitate to fling another pot, and as their cries of 'More!' kept coming, I threw all the bowls, pans and cans onto the pavement. My neighbors continued to show their approval, and I was very happy to give them pleasure. But my supply was used up and they kept yelling, 'More!' I hurried straight into the kitchen and fetched the earthen plates, which made an even jollier show when they broke, and so I ran back and forth, brought one plate after another, as I could reach them in a row on the pot shelf, and since the neighbors were still not satisfied, I flung all the utensils I could get my hands on into ruin."[34]

Bunzlau doll utensils in a brass
wire sack, 1920's. Height of the
pitcher 32 cm.

Color plate
Coffee service. Above left:
porcelain, white with gold, circa
1850. Height of the coffeepot 6 cm.
Above right: tin, circa 1890.
Height of the pot 6.5 cm.

Lower left: plastic, circa 1950.
Advertisement for Günzburger
Coffee. Height of the pot 4 cm.
Lower right: porcelain, 1920's.
Height of the pot 6 cm.

So the small earthen utensils intended for children
were sold at pot markets along with the big ones.

Another childhood reminiscence made clear the
popularity of these utensils. Bogumil Goltz (born 1801)
wrote in his autobiography in 1854:

"The Warsaw potter's art then produced, as even
now, the dearest children's toys, wondrously cute little
pots and bowls, pitchers and stewpots on three legs. I
could get wrapped up in the wonders of these glazed
duodecimo utensils for hours and think about how
they must have been made. For I had no concept of
potter's clay on a potter's wheel, or of how it was baked
and changed into an artificial mass of stone with
glazing and coloring.

"Like all children, we also liked best to play with earthen toys, and whoever felt at all bored or ignored in mixed company usually excused himself by saying, 'Well then, I'll take my pots and bowls and go home'."[35]

Every pottery region had its own typical forms and colors, which were also used for children's utensils. The Bunzlau utensils were of more than regional importance in doll kitchens as in big kitchens. They were sold, also in small sizes, in the typical patterns of the times. Even in the 1930's one could buy glazed pots and pitchers in solid browns and blues or mixed shades of brown and blue, packed in little brass wire bags (see the doll kitchen on page 47).

The eggshell-colored utensils with light blue, brown or rust-red cut checkerboard patterns from Znaim in Moravia were similarly widespread. These strictly geometric patterns gradually replaced the traditional carved names, lines and flowers after 1900 (see the doll kitchens on pages 55 and 61).

Earthen doll utensils, made circa 1905, light blue and white. Height of the pitcher 7.8 cm. Probably from Znaim, Moravia.

Glazed earthen pots from the
National Street Collections of the
Winter Welfare Agency, 1940 and
1942. Height of the pitcher at right
rear, 4 cm.

Another group of small earthen utensils is often found in old doll kitchens: cans, pitchers, cups, baking pans, bowls and roasting pans in lime green, cream, brown and rust red. They were first sold with WHW trademarks at the National Street Collections at Magdeburg-Anhalt in 1940 and South Hannover in 1942 (see p. 87).[36]

Organizing a list of regional earthenware centers in terms of time and area is not easy, especially as earthenware was produced in traditional ways over long periods of time, and the potters regularly took popular patterns from each other. The criteria for identifying doll pots are the same as for large utensils; therefore literature about pottery in general can also apply to small pots.

"Majolika utensils with nickel
trim", 1900. Märklin 1, page 227.

Fayence or Majolika

Fayence is not incrusted, that means the texture is porous as is that of earthenware, but the glazing, on the other hand, is covering, and painting can be done on the glaze and is burned in through further burning processes. The fragile Fayence utensils survived only seldom in doll kitchens. The Märklin Brothers of Göppingen had in their 1900 catalog "Tableware from Majolika with nickel trim, very excellent finish with finest flower decor", suitable for similarly decorated enamel stoves. (This may actually be pottery which, when painted, has sometimes been called Majolika, i. e., from Majorca.)

Pottery

Children's coffeepot, circa 1860. Pottery, light blue transfer print, height 11 cm.

Pottery utensils resemble porcelain in appearance. But the non-translucent texture is, like that of earthenware and Fayence, ont incrusted. Patterns are usually applied to the unbaked surfaces by transfer printing or with the use of adhesive pictures. These cheap goods were always very popular in doll kitchens.

Parts of a doll table service, circa
1860. Pottery with green print.
Height of the tureen 9 cm.

Parts of a children's table service,
circa 1914. Pottery decorated with
adhesive pictures. Height of the
tureen 11.5 cm.

Stoneware

Stoneware is incrusted like porcelain, but made of chemically unpurified raw materials and therefore gray, brownish or yellowish. In doll kitchens one usually finds gray cooking pots, painted in cobalt blue and finished with salt glaze. They were always made, like earthenware, by the manufacturers of large goods. The most important centers pf production are to this day around Siegburg and in Höhr-Grenzhausen in the Land of the Pitcher Bakers.

Stoneware pots, circa 1880-1910.
Gray with blue decor, salt glaze,
probably from Höhr-Grenzhausen.
Height of the bench 9.5 cm.

Porcelain

In general, porcelain did not yet play any great role in the kitchen during the first half of the Nineteenth Century. Earthenware utensils were used most often, along with copper, tin and iron. Until the middle of the century, the kitchen was usually still influenced by the smoke and soot from the open hearth. Only with closed stoves could a friendly, bright atmosphere be established in the kitchen. The love for white porcelain seems to have spread in this era, encouraged by urbanization and improvement of the standard of living among the bourgeoisie. White pitchers with measures marked on them, canisters, spice racks with inscriptions, oil and vinegar bottles, salt and flour containers, spoon and implement racks with cooking spoon, whisk, ladle, rolling pin, souffle/and pudding forms and bread boards of porcelain were put on display in 1883 in a "porcelain kitchen" that the Schumann Porcelain Factory set up as part of the Hygienic Exhibition in Berlin.[37] Into the present century porcelain was the "pride of the housewife", and in some ways it took over the role that copper had played in the elegant kitchens of the Eighteenth Century.

It seems that many utensils for doll kitchens were already made of porcelain at the beginning of the Nineteenth Century. In any case, Hieronimus Bestelmeier's 1803 catalog shows a kitchen furnished with porcelain as well as tin (No. 400), and No. 541 is "a food container (cabinet with fly screening; author's note) with various utensils of tin and porcelain in it."

Like the big kitchen, the doll kitchen was richly furnished with porcelain only in the latter half, in fact the last third of the Nineteenth Century.

Chicken (basket) for the doll kitchen, circa 1900. Painted porcelain. Height 3 cm.

Kitchen furnished with porcelain (for adults). Displayed at the 1883 Hygienic Exhibition in Berlin. (Illustrated Women's Magazine, 1883, p. 22)

Children's coffee service,
porcelain, 1922. Photograph.

Kitchen utensils of porcelain,
onion pattern, circa 1875. Original
box 35.5 x 47.5 cm. (Historical
Museum, Frankfurt/Main)

Porcelain with painting under the glaze enjoyed great popularity. The "onion pattern", developed in Meissen in 1739 after Asian models, was a pattern copied by many manufacturers. Similarly treasured, if not even more so, was the "strawflower pattern" used in doll kitchens. It came most often from Thuringian manufacturers, such as Huttensteinach or Rauenstein, but the Franconian firm of Tettau also had it in production. Doll kitchen utensils generally bear no trade marks!

Parts of a doll table service, circa 1870. Porcelain with blue pattern under the glaze. Height of the tureen 6 cm.

Less noticeable but not at all rare are the colorfully glazed porcelain pitchers and mugs of barely 2 to 7 cm height found in doll kitchens in the last decades of the Nineteenth Century and the first decades of the Twentieth. The relief work on the outside shows flower patterns as well as tavern scenes; tree-bark designs with owls and other birds or simple lengthwise grooves were also common. The brown colors outweigh the blues and greens. The most popular products among this colorfully glazed porcelain are probably the beer steins that have been in doll kitchens for many decades in this century. In Austria they were produced by the firm of Steidl (p. 165).

Pitchers, circa 1890. Porcelain, relief designs, glazed brown. Height of the largest pitcher 6 cm.

Doll table service of porcelain,
circa 1913. (Borho, Baden-Baden)

Parts of a doll table service, circa 1900. Porcelain, decorated in light blue and gold. Height of the tureen 7.1 cm.

Parts of a doll table service, 1920's. Porcelain with lime green adhesive pictures, tomato-red painting. Height of the tureen 6.8 cm.

Parts of a doll table service, 1930's. Porcelain, adhesive pictures, height of the tureen 6.8 cm.

Glass

Delicate glass items blown over the flame belonged in doll houses and kitchens. In doll kitchens, lower-quality bottles and water glasses are most commonly found, as well as the indestructible, richly decorated moulded glass that was already used as low-priced goods in adult households at the beginning of the Nineteenth Century.

Cup, sugarbowl and salt-pepper holder for the doll kitchen, circa 1870. Light blue moulded glass. Height of the bowl 5.6 cm.

Wood

There have always been wooden utensils in the doll kitchen as in the big kitchen. Cooking spoons, rolling pins and cutting boards exist to this day. Of course the competition from plastic has grown great over the years.

To a certain extent, large and small wooden kitchen utensils have been produced in all regions rich in wood, but the toy centers around Sonnenberg and in the Erzgebirge were the most important producers of "toy-box wares" for the Nuremberg trade. Complete household furnishings in miniature—turned in wood—were packed in splint boxes and wooden chests. For example, the Sonnenberg Pattern Book of 1831 offers five wooden boxes with small turned "House-wares" (No. 414-418), as well as "Häfelein" pots, small, ditto, medium, ditto, large" and, likewise in three different sizes, "Hand baskets, "Stützlein" baskets (with legs), back baskets, butter tubs, buckets, firkins, plates" of turned wood with painted red and green stripes (No. 368-391). Even richer are the offerings of the Waldkirchen Pattern Book of around 1850. A wealth of turned wooden utensils, (Plates 8, 10, 30, 86) as well as implements such as ironing boards, butter tubs, whisks, cooking spoons, rolling pins, baskets, buckets in various sizes, brooms and brushes (Plate 87) were available. Along with these wooden utensils and containers "small as well as big", there were also many wooden things made for doll kitchens that in reality consisted of very different materials. The Waldkirchen Pattern book of around 1850 offered, for example, painted plates that resembled porcelain, bearing fruit, fish, lobster, cake, bread and cheese (Plate 66). There were also splint boxes with imitations of metal household goods, such as brass-painted candlesticks and mortars, plus pots and pans in copper tones. Even

wineglasses, pitchers, three-legged pans for the open fire and "sow-snouts" (earthenware roasting pans) were turned of wood. Especially charming are the wooden services that show delicate "porcelain painting" on white or bone backgrounds (Plate 66). A Nuremberg pattern book from the years around 1850-60 displays numerous cardboard boxes, covered with colored paper, that contain beautiful coffee, tea and table services that look like porcelain or Fayence. They are actually made of finely turned wood and painted (pp. 34-35).

The iron rings of the "Böttcherware" were not painted, but imitated in tin. The turner held a tin rod of the correct width against the quickly rotating piece of work. The metal rubbed off formed a thin film around the turned utensil and made a silver stripe.[38] This finally became a decorative element in its own right and was also applied to turned coffeepots and other vessels.

Doll dishes, 1850-60. Turned wood, painted, packed in cardboard boxes. (Nuremberg Pattern Books, page 69)

Wooden articles, second half of the
Nineteenth Century. Height of the
chopping block 10.8 cm.

The complete sets of housewares in wood, imitating other materials, gradually disappeared from the pattern books and catalogs, for the sheet metal utensils made in Nuremberg, which were similarly priced, and which could also be used on the new alcohol stoves with open flames, pushed them aside.

Doll kitchen utensils of wood, 1850-60. (Nuremberg Pattern Books, p. 73)

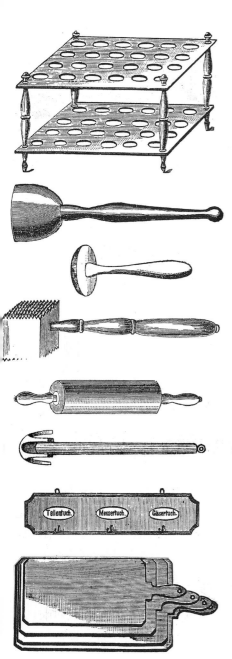

Left: wooden utensils, circa 1890-1930. Egg rack, potato masher, pea masher, meat mallet, rolling pin, whisk, towel rack, cutting boards; above: salt shaker (with "Flour" lettered on it!) and spice cabinet.

Naturally, individual wooden implements from mother's kitchen were still sold in small sizes. For example, a report on the Vienna World's Fair of 1873 says: "In the Thuringian Forest wood is carved into: spice cabinets, salt and flour barrels etc."[39] A very typical range was produced by the wooden goods factory of Arthur Faber in Bietigheim, Württemberg. In a 1910 catalog of kitchens household goods and furniture they offered a manifold selection of "fine children's toys" including the following articles: "beefsteak mallet, maple; bellows, beech; wooden knife, butter knife, maple; spice chest, yellow wood polished; pickle slicer, meat mallet, potato masher, cutting boards, spoon rack, flour scoop, spoons, maple; knife box, yellow wood polished; knife sharpener, whisk, maple; salt mortar, maple polished; salt-cellar polished; hair sieve, Christmas cookie cutters with six shapes; cleaning box, beech; tablecloth (wood strips), polished; meat board, maple; noodle board, cutting board, beech; rolling pin, maple; serving table, oak; laundry drying rack, beech;" plus utensils such as slicing knives, cleavers, graters and skimmers.

In addition, Faber shipped the most popular items from this collection "in strong boxes with sliding lids"; there were three assortments, with 20, 25 and 35 items. The cheapest cost 5.80, the most expensive 19 Marks.

Implements and machines from the doll kitchen

Upper left, to circa 1870 On the bench, from left to right: slicing knife, oil lamp, taper box, sugar breaker, coffee machine with filter, shopkeeper's scales. Below, from left to right: mortar, chafing dish, two handmade copper pots, coffee roaster. (Historical Museum, Frankfurt/Main) Height of the bench 11.2 cm.

Upper right: last third of the Nineteenth Century On the table, from left to right: grinding machine, potato press (Sig. Blömer, Frankfurt/Main), Bread-slicing machine (front), store scale (Sig. Blömer, Frankfurt/Main). Below, from left to right: Ice cream machine (Historical Museum, Frankfurt/Main), salad spinner or egg-cooking basket, window bucket (Historical Museum, Frankfurt/Main), cream whipper (Sig. Blömer, Frankfurt/Main), knife-sharpening bench (Historical Museum, Frankfurt/Main), petroleum lamp. Height of the table 12.5 cm.

Lower left: beginning of the 20th Century On the table, from left to right: grinding machine, knife-sharpening machine (Sig. Blömer, Frankfurt/Main), butter machine, meat grinder (Historical Museum, Frankfurt/Main).
Below, from left to right: food carrier (Historical Museum, Frankfurt/Main), bread basket, table scale. Height of the table 10.5 cm.

Lower right: 1920's On the sink, from left to right: whetstone, store scale, egg-shaped tea ball, thermos bottle, bread-slicing machine. Below, from left to right: icebox, "Kitchen Wonder". Height of the table 11.6 cm.

"Children's cookstove, lacquered and bronzed", end of the 19th Century, with cast iron and enameled sheet metal utensils. Offered by the Holler'sche Carlshütte AG near Rendsburg. Color lithograph. (Sig. B. ten Kate, Amsterdam)

Copper

In the time of Louis XVI (born 1754), "complete children's kitchen furnishings of red copper" are said to have been made.[40] The doll kitchens of the first half of the Nineteenth Century were also, in part, richly equipped with copper.

Copper water kettles, cooking pots and decorative baking forms were kept in production longest. The Viennese toy dealer Anton C. Niessner still had them in his 1910 catalog, and even Märklin of Göppingen, the toy manufacturer who used predominantly sheet metal, still offered copper fish, grapes, crabs, clams and melons in 1895. In 1909 the same items were available only in "sheet tin, stamped and plated, with bordered rims". For most stoves Märklin still offered optional copper pots as late as 1919.

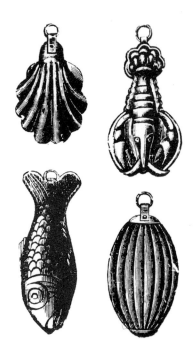

Copper forms. (Forms that can be found in doll kitchens during the entire Nineteenth Century)

Copper utensils: inverted fat-bellied milk cans, cooking pots, water kettles, baking forms. Brass utensils: basket-type pail and chafing dish at left front, fire shield near the hearth, in a doll kitchen, circa 1800 (see page 21).

Brass

In the lavishly appointed, rare doll kitchens of the Eighteenth Century, many utensils were made of brass or even silver. In 1803 Hieronimus Bestelmeier offered a systematic list of his wares, including "2. Play and useful items for boys and girls", among which was "brass housewares 36 kr.", as well as wooden, tin and sheet metal utensils.

The old artistically pierced implements of sheet brass that survived from the Eighteenth Century: warming stoves, pierced baskets, coal pans and buckets, can be seen in the doll kitchen of the Historical Museum in Frankfurt (p. 21).

Pans with handles, casseroles (deep pans with lids), ladles, beating bowls with rounded bottoms, marmelade bowls with flat ones and scales of brass remained as doll kitchen furnishings into the 1920's, and in a few cases into the Thirties.

Among the cast brass goods were also implements that could stay in production until well into this century, such as dough wheels. "Brass children's mortars" and "brass irons for children" were offered in Wathner's Iron Goods Connoisseur of 1885,[41] and Anton C. Niessner of Vienna still had them in stock in 1910.

Mortar, circa 1870. Brass. Height without pestle 3.8 cm.

Sheet metal utensils, second half of the 19th Century. (From left to right) Rear: plates, lithographed and plain. Center: pitcher and cream whipper, sheet tin; coffee mill lacquered in mahogany color; flour canister, lithographed. Front: buckets, washtub and pitcher painted in wood colors; spice cabinet painted mahogany, bronzed. Height of the sideboard 29 cm.

Washing utensils, 1930's. Wood.
Length of the brush 33 cm.

Tin

Decorated tin utensils and implements have been made and sold in miniature form to this day—with a short interruption for World War II and the first postwar years. But big kitchen utensils of tin, the smooth plates, bowls, carving sets and canisters, were pushed aside at first by sheet metal housewares in the latter half of the Nineteenth Century, and finally by porcelain as well. But to this day they fill the open shelves of doll kitchens.

Old toy catalogs such as that of Bestelmeier do not specify but simply offer "a tin set of housewares, 1 fl. 12 kr.[42] The people who ordered them seemed to know what to expect. In addition, one of the kitchens was offered "with utensils of tin and porcelain (No. 400) and another equipped only with tin (No. 1012).

Canister, end of the 18th Century. Tin, handmade. In one upper corner a baffle has been built in to slow the flow of the grain or other substance when pouring out. Height 6 cm. (Historical Museum, Frankfurt/Main)

Tin doll utensils, 1850-60.
(Nuremberg Pattern books, p. 71)

The most important centers of production were in the Nuremberg-Fürth area and in Saxony (for example Freiburg, Gotha and Dresden), but Berlin firms also made tin doll housewares. At the 1844 General German Trade Exhibition in Berlin, "Joh. Andr. Weigmann, master tin caster", presented: "in addition to enema injectors, also three chests of children's housewares for 12 kr., 42 kr. and 2 Gld. per chest," and "G. Söhlke, toy manufacturer from Berlin: various children's toys, mainly made of tin. All sorts of pretty housewares were among them. . . ."[43] In Diessen on the Ammersee, doll utensils are still produced today, partly in old forms.

Doll stove, circa 1880. Sheet iron;
doors, poker, frame and lion's feet
of brass; for alcohol; copper
cooking utensils with tall cocoa
pot. F. & R. Fischer, Göppingen.
Height without stovepipe 16 cm.

Doll kitchen, circa 1910. Housing
(36cm high, 59/42.5 wide, 30 deep)
of wood with white and blue
tiling. Stove: sheet metal, steel
blue, nickel-plated parts. Utensils:
white enamel, wood.

Sheet Metal

Cooking pots, basins, funnels, cans and many other items for the kitchen were made at first by sheet-metal workers, and around the middle of the Nineteenth Century they were finally produced by machine in large quantities. Soldered joints, rivets and folds on cut sheet metal pieces and specially made bases are characteristic marks of the older pieces.

A real industry for sheet metal goods could develop only around the middle of the Nineteenth Century, a few decades after the invention of "pressing on the lathe", for the formerly customary shaping of vessels with the chasing hammer was laborious. But with the new technique, objects could be pressed quickly and thus cheaply over a hardwood form. Forms that curned inward, of course, had to be produced in parts and finally soldered, riveted or mortised together. Seamless tall vessels that were formed of one piece of relatively thick sheet iron on the draw press came from the last decades of the Nineteenth Century as well as our own. The sheet metal utensils were cheap, but they became unsightly very quickly, discolored foods and gave them a metallic taste. Therefore they were tinplated. In 1822 the firm of Justus Assmann was already displaying such tinplated sheet metal utensils successfully at the German Trade Exhibition in Berlin, as "Neuwied Health Utensils".[44]

For the children there were also sheet metal utensils, at first made by artisans and likewise tinplated, but also often painted. Once again evidence is found in Hieronimus Bestelmeier's 1803 catalog, in his systematic listing under "2. Play and useful things for boys and girls". There we find: "A chest of sheet metal housewares, lacquered in Wedgewood style, 36 kr." In 1843 a Merseburg toymaker offered "kitchen utensils of sheet metal, in boxes of 12 to 36 pieces, the box from 4

Child's pitcher, circa 1880, nickel plated sheet metal. Height 10.5 cm. Märklin Brothers, Göppinge

Sgr. to 12 Sgr.'',[45] and in the following year a master sheet-metal worker of Zerbst, Wilhelm Kohl, displayed at the German Trade Exhibition in Berlin "thirteen boxes with painted toys of various kinds, including a table setting, coffee service, etc.''[46] The "Toy Factory of C. P. Dietrich, Ludwigsburg in Württemberg'' was also represented with "two boxes with kitchen utensils for 20 Kr. and 1 fl. 13 Kr.'' In another place it was stated that the firm had displayed "in addition to other toys, numerous kitchen utensils of sheet metal''.[47]

Doll housewares of sheet metal, 1898. Advertisement of the mail-order house of Mey & Edlich, Leipzig.

Pitcher with cup, circa 1800.
Porcelain, green painted trim.
Height 6.6 cm. (Sig. G. Ullmann,
Munich)

"Cookstove with complete kitchen furnishing" (G. Söhlke, Berlin, 1893)

Stove, 1899, sheet metal with nickel parts. (Herz & Ehrlich, Breslau)

Sheet metal toy factories, such as those of Ludwig Lutz of Ellwangen, Rock & Graner of Biberach and Märklin of Göppingen, also produced sheet metal furnishings for doll kitchens. The Märklin couple, in fact, began their toy production in 1859 with furnishings for doll kitchens, and their sons, who took over the business in 1888 as "Märklin Brothers" after their parents' death, produced doll kitchen furnishings and cookstoves at first.[48] As late as the Twenties these toys were a significant part of Märklin production. The Austrian wholesale catalog, "Wathner's Iron and Iron Goods Guide", of 1885 lists under point II: "Pressed and Folded Sheet Metal Utensils" 171 objects for the big kitchen and then continues: "After the described sheet metal utensils, children's toys are also sold in sets or individually; these are limited, though, to cooking utensils such as: pots, casseroles, pans, roasting pans, baking forms, kettles etc., and are often assorted including the smallest objects, for example graters, cream whippers, lids, cups, funnels, spoons, candlesticks etc."[49]

It is almost self-evident that a great array of sheet metal utensils appeared in the toy catalogs of the end of the Nineteenth Century and the first ten or twenty years of the Twentieth. The geometric or figured patterns of the plates and bowls of these times are particularly charming.

Sheet metal utensils for children were often painted. As with turned wood utensils, attempts were made to imitate other, more expensive materials. One need only recall the aforementioned "house furnishings in Wedgewood style" in Bestelmeier's 1803 catalog. In the pattern book of G. Striebel of Biberach, circa 1850, sheet metal cans, cups and pitchers are illustrated that imitate Fayence or porcelain utensils of the time in form and finish. Baskets, buckets, pitchers and the like of tinplate were, like soft wood articles, often painted in ring patterns. Examples are found from Lutz of Ellwangen. For the Biberach toy manufacturers, Rock & Graner, mahogany-colored kitchen goods were typical of the 1870's: bread boxes, spice cabinets, coffee mills and water buckets. Brass-colored and bronzed canisters, with various types of lettering in different eras, were widely known.

With the coming of enamel, it was natural that the sheet metal utensils would be painted, sprayed or dipped in the manner of enamel, with clouded, speckled and other patterns. Light blue was the most widely imitated color.

Of course sheet metal toys for the doll kitchen were not just oainted and lacquered. As in the entire area, the color printing of sheet metal finally took the place of hand-painting. Since the color was not well transferred by the pressure of the lithographic stone on the metal, the so-called "transfer printing" process was used, in which the picture or lettering was first printed on the "transfer paper" and from that onto the metal. The simpler offest lithography was generally accepted from the Eighties on. Only from this time on did the doll kitchen have an astonishing richness of printed objects, especially plates, flour and salt boxes and canisters. Tee services brightly printed in Chinese decor or with children's motifs can be found in every toy catalog until the 1930's. After World War II they were produced anew and then gradually replaced by plastic.

Canisters, circa 1890. Sheet metal, gold-bronzed with black lettering

Doll utensils, sheet metal, 1912. (Stukenbrok, Einbeck, p. 310)

Parts of a doll tea service, 1949. Sheet metal in Twenties-Thirties style. Printed red-blue-black Chinese motifs on yellow-beige background. Height of the teapot 6.4 cm.

Doll housewares. Thirties. Sheet
iron, lacqured in speckled light
blue and white, sewn onto the box
(45.5 x 31 cm). France (M. J.,
Paris.)

Enamel

Utensils made of sheet metal were enameled in great quantities only in the last third of the Nineteenth Century. The first cast iron vessels in Germany are said to have been enameled in 1785 by the Gräflich Einsiedelschen Eisenwerk of Lauchhammer, in the district of Liebenwerda, one of the oldest German iron works. In 1844 this same firm and four others sent to the German Trade Exhibition in Berlin a variety of cast iron, white-enameled cooking and roasting vessels. In addition, the show report cited sheet metal utensils covered with enamel as new products.[50] But attempts to enamel sheet metal had already been made ten to twenty years earlier.

Doll utensils were not yet mentioned then. Wathner's Iron and Iron Goods Guide of 1885 states generally in the chapter on "Enameled and Tinplated Goods": "Much more uesful than tinplating is enameling, and most of the vessels and implements that were formerly

Parts of a doll coffee service, circa 1900. Sheet iron, enameled in white with forget-me-nots. Height of the pitcher 4 cm.

made of copper, tinplate or clay are now produced in sheet metal or cast iron and enameled, and take on a very nice appearance, especially as objects that are enameled in white inside and out look like porcelain utensils."

The cast iron goods were generally either enameled or sold raw; the "pressed and folded sheet metal utensils", on the other hand, were either tinplated or enameled. . . ." and in addition there are the finer types of utensils. . . and many others can be mentioned that are enameled in white inside and out, sometimes also with decorative designs. In addition to the described sheet metal utensils, there are also children's toys sold in sets or individually, but these are limited solely to cooking utensils . . ."[51] The enameled sheet metal utensils were the same types as were tinplated.

In older doll utensils, the edges soldered over each other, the folded seams, the rivets of spouts and handles, the individually attached bottoms still show through the enamel. The old work processes, to be sure, were still used in this century along with production on draw presses,[52] so that no clear possibilities of dating can be derived from this production method. The colors, too, only hint at the age to a limited degree. The dark glossy blue—though it was still used in the Thirties—is the older blue, while the light blue, the "new blue", was in fashion at the end of the Nineteenth Century.[53]

Enameled doll coffee service, 1899.
(Herz & Ehrlich, Breslau)

Emaillirte Puppen-Caffee-Service,
aussen hellblau, innen weiss.

H.&E.

H.&E.

Berliner Form.

*Porzellan-*Form.

Parts of a doll coffee service from a doll kitchen of 1910-20. Sheet iron, enameled, speckled gray and white. Height of the pot 12 cm.

Utensils from "prime sheet metal-enamel, outside brown, inside white", with lacquered iron stove (Herz & Ehrlich, Breslau, 1899)

Enameled doll utensils, unfortunately, bear trade marks only rarely; thus it is almost never possible to ascribe them to individual firms. Baumann of Amberg and Bing of Nuremberg definitely produced doll utensils,[54] and presumably other enameling works did the same. There seems in any case to have been much demand, for most toy catalogs showed enameled doll utensils among their girls' toys at the end of the last century:

G. Söhlke's Successors, Berlin (1892): "Kitchen utensils, enameled iron at 1.50, 2.50, 4, 5, 6 and 7.50 Marks" and "new cookstoves with blue and white enameled utensils".

Märklin Brothers, Göppingen (1895): "Cooking utensils enameled white inside, brown outside; enameled coffee service, white with blue and gold or pink and gold decoration".

Enamel utensils, circa 1920, sheet metal enameled light blue outside, white inside; on a white enameled sheet metal stove with stamped brass legs, 1920's. Height (without stovepipe) 14 cm.

166

Child's coffeepot, last third of the 19th Century. Sheet iron, enameled white inside, copper brown outside. Base attached, spout and handle riveted. Height 14 cm.

F. & R. Fischer, Göppingen (1896): Stoves also optionally "with enamel utensils."

Herz & Ehrlich, Breslau (1899): ..."the utensils are first-class sheet metal-enamel, brown outside, white inside. . . Cooking utensils are enameled white with blue stripes. Enameled doll coffee service, light blue outside, white inside, in Berlin form and porcelain form."

In the first ten years of the Twentieth Century, enamel utensils were probably the most popular kind of doll utensils. They toy dealer Anton C. Niessner of Vienna, in 1910, offered utensils in "granite blue, white inside and out, blue and white inside, copper brown outside and white inside, white speckled with light blue."

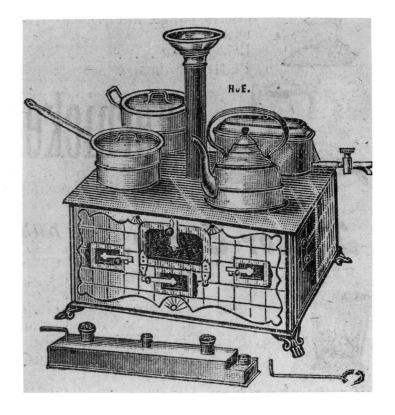

"Children's cookstove, enameled in fine white tile pattern, with gold trimming, ooking utensils enameled white with blue stripes." (Herz & Ehrlich, Breslau, 1899)

167

Aluminum

Aluminum is often regarded by collectors as a modern metal that has no place in older doll kitchens. But aluminum was discovered in 1827 and produced in factories in Paris by 1855. Mayer's Conversational Lexicon of 1896 says that there are many uses for this material because it does not rust and is easy to work. Anong other things, cooking utensils have been made of it for many years.

No date has yet been set for the first aluminum pots for dolls. In any case, the Märklin Brothers of Göppingen had put aluminum toy cooking utensils into production by 1909 at the latest. In their catalog of that year they offer three stoves in different sizes, optionally equipped with aluminum utensils, aluminum services, and also a set of aluminum pots.[55]

In toy catalogs after 1910, aluminum utensils had already gained a good place along with enamel and porcelain. The prices at that time were still about the same. Aluminum utensils, "with fine matt finish similar to silver" were offered not only by Märklin but also by such dealers as Borho of Baden-Baden circa 1913, not only for the stove but also as coffee and punch services. Sieves, ladles, tureens, plates and silverware, baking spatulas, bowls, buckets, small tables, salt and flour canisters, containers for sand, soda or soap, and in the end any and all housewares for the doll kitchen were made of aluminum (see page 68).

After World War II aluminum reached a second high point in the doll kitchen, but was then pushed out of the picture by a new material, plastic. The firm of Masteler & Killmann, of Kettwig on the Ruhr, a metal goods factory that offered many aluminum items, also produced "aluminum children's utensils" in 1914, and was not alone. The metal goods factory of Hesmer & Möllhoff, of Bärenstein, Westphalia, put out a special

Electro-stove with aluminum utensils, circa 1950. Height 13 cm. (Historical Museum, Frankfurt/Main)

Aluminum doll stove with aluminum utensils, circa 1950. Height 17.5 cm.

Aluminum cooking utensils for
the doll kitchen, circa 1913.
(Borho, Baden-Baden)

calatog of ''aluminum toys'' in 1927. Twenty-nine
pages of the complete housewares of the time in
''polished or burnished'' and ''matt silver stained''
form were offered individually and in a variety of sets.

Doll coffee service, pure
aluminum, circa 1913. (Borho,
Baden-Baden)

Plastics

In the Twenties and Thirties, lampshades as well as some small utensils were made of celluloid, and small and larger doll dishes were produced in bakelite, in green, brown and red shades. The real boom for utensils of synthetic materials took place in the era after World War II. At first polystyrene was the most popular material for small advertising promotions, such as the well-known ivory-colored figurines of a margarine manufacturer and the pastel-tinted coffee services that the "Korona" coffee company packed in their coffee-bean mixtures, one piece at a time, for people to collect. Packets of "Günzburger Instant Coffee" also contained business-promoting little cups, plates and coffeepots. These utensils existed in, among others, transparent reds and blues with gold rims, as well as in cream with adhesive pictures in the form of old tin utensils. From the Fifties and Sixties on, the toy trade offered modern forms, chiefly in pastel shades at first.

Parts of a doll tea service, circa 1940. Bakelite, mottled green, England. Height of the pitcher 5.5 cm.

Wire Goods

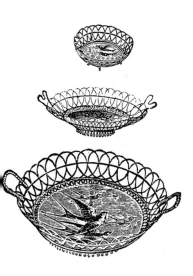

Fruit and garbage baskets, circa 1890. Wire and pottery.

Unspectacular but particularly charming items in doll kitchens are the small wire utensils, the almost ageless whipped-cream whisks, spoon holders, onion baskets and nets, cake racks and hot-pot mats, salad spinners or egg-cooking baskets, mashers, bottle and glass carriers, silverware baskets, mesh food covers and hair sieves. Small wire baskets with nicely painted pottery bottoms are often found in doll kitchens too. In big households they were called "fruit baskets" if their diameter was 20 to 30 cm and "garbage baskets" or "potato-peel baskets" with a diameter of about ten cm.

Most wire goods represent typical utensils of the Nineteenth and the first decade of the Twentieth Century, but many of them, such as the cake rack and the hot-pot mat, still exist.

Appliances and Machines

Many of the old kitchen implements in the doll kitchens, such as the grating and chopping machines, are hand-powered forerunners of our present-day electric appliances. Other devices such as sugar breakers and knife sharpeners have not existed for many years. Their purposes and functions are comprehensible today only if we have knowledge of earlier times. So we shall describe some especially interesting and significant rare doll kitchen implements in the following text.

Bread-slicing machines
existed at least in the 1870's. they had holding boards set at right angles to each other and a curved cutting knife on a handle that increased the pressure when cutting. The small doll bread slicer functioned in the same way but could cut only marzipan. Like the big bread slicer, the small one developed into a round blade that found acceptance only in the Twenties but has remained the basic principle of the electric bread slicer or general kitchen slicer.

Glass butter machines
came on the market around 1900, at which time making butter in a big butter churn at home was already a thing of the past. Butter was then bought in special milk-butter-cheese shops or from farmers at the market.

The household butter machine arose at that time as a result of a need for fresh, healthy, unadulterated foods, a need that we are familiar with today. "Costly butter for nothing and excellently tasting buttermilk (recommended by doctors) can be made by every practical housewife from the cream of the daily milk," said an advertisement of 1901.

The glass container was filled with cream and firmly closed with the screw-on stopper to which the wooden whisk is attached. a gear on the hand crank made for relatively high speed.

Ice cream machines

for the doll kitchen were thick-walled insulated sheet-metal containers to be filled with a mixture of ice and salt into which the actual ice cream container could be inserted. In Holzbottich machines the ice cream could be stirred during the freezing process.

Meat grinders

"These machines chop bits of meat thoroughly and can be used just as well to chop spinach, cabbage, etc." (Household catalog, 1895)

The meat grinder has not been replaced completely by electric appliances to this day. It can still be bought, even as a small doll-kitchen grinder made of aluminum.

Chafing dishes

are perforated iron containers with a wooden handle, used to warm foods or keep them warm (p. 14).

Coffee roasters

In the Nineteenth Century coffee was usually roasted at home; only as of about 1900 was it usually bought in finished form. The raw beans were yellowish, gray or greenish, depending on their place of origin. The greenish types were regarded as the best. Since the other types were often colored green (with copper vitriol!), the beans had to be washed thoroughly several times before roasting. After drying they were put into a roaster, which was made to fit any source of heat. All coffee roasters were based on the same principle, namely that of keeping the beans in motion inside a closed container to roast evenly over the fire. Stopping the air from escaping was supposed to hinder the escape of what produced the aroma.

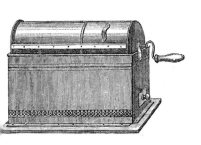

Various coffee roasters, last third of the 19th Century.

Cooking boxes
had many names and many inventors. They first
appeared officially at the Paris World's Fair in 1867. In
a household lexicon of 1884 they were referred to as
Sorensen self-acting kitchens, and in a 1904 cookbook
they were called Goehde's self-cooking apparatus.

Many a housewife had already used the principle
long before. To save energy and time, she put the
precooked food, such as rice, into a featherbed and
pressed a pillow firmly around it. After several hours
the rice was fully cooked. The self-cookers or cooking
boxes work in just the same way: a hot precooked food
is put into a firmly closed pot that is insulated and kept
at a temperature that is sufficient to finish the cooking
process.

As an ideal complement to the gas stove, the cooking
box was built into the famous Frankfurt Reform
Kitchens of the 1920's as standard equipment.

The Märklin Brothers offered a functioning "Youth
Self-Cooker" for the doll kitchen in 1909.

Coffeepot with filter, 1896.

"Youth Self-Cooker", 1909.
(Märklin 4, p. 107)

Cooking box, circa 1910. Sheet
metal, printed, with two pots to b
put in it. Height 4 cm. (Historica
Museum, Frankfurt/Main)

Knife-sharpening bench, circa 1880. Wood.

Milk warmer, 1880.

Grinding machine, 1896.

Knife-sharpening benches

were for the simple iron knives, found in most households, that could be sharpened well but rusted easily. They were simply wooden benches covered with emery cloth, with a leather strop "for finer honing of the steel".

The knife-sharpening machine removed rust with two rollers set close beside each other and turned by a hand crank.

The milk warmer

was a practical invemtion for mothers who had to warm milk for their babies at night or during the day when the fire was banked. It was built like a tea warmer.

Oil lamps

are not as old as people like to think. America's worldwide sales only became significant as of 1859; thus oil lamps and cookers can have played no role in Europe before 1860.

Grinding machines

are among the most popular appliances in the doll kitchens of the last hundred years. The cast iron body always conforms to the fashion of the times.

Egg-white beaters

were simple wire whisks, still known today, or implements in the form of high sheet-metal cans in which a perforated plate of sheet metal (sometimes several) on a long handle could be moved up and down quickly.

Taper boxes

were useful containers for wax tapers whose ends projected through the hole in the lid, and which could be taken out when needed. The wax taper itself was a popular source of light, especially in the 16th to 19th Centuries, a long, very thin rolled candle.

Sugar breakers

Only after 1800 did the previously imported cane sugar get any competition from home-grown beet sugar in Europe. At that time sugar lost the character of a costly kitchen seasoning. Sugar was sold in the form of candy, loaves and "sand" or granulated sugar. Loaves of sugar were generally made of a mediocre quality of beet sugar. The hard loaves could be broken to pieces with a hammer while wrapped in a cloth. But there were also nicely made, sturdy sugar breakers. Such a tool can be seen in the doll kitchens of 1860-70 on page 21. The loaf, held fast between the board and the cutter, is broken by means of a long handle.

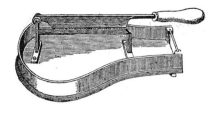

Sugar breaker, 1880.

Egg-white beater, 1896.

Baking utensils from the doll kitchen, second half of the Nineteenth Century. Height of the table 24.5 cm.

Breadbox, circa 1880.

Washing Machines

Until the days of electric washers after World War II, simple wooden washing machines could be bought for children. There were washtubs on frames, washboards and wringers, and clothespins and a pulley for the washline were usually available too. These simple devices were also used in the big household. Some people washed as best they could in the kitchen; others had a laundry room to use, usually located in an outside shed in the country or in their cellar in the city. The laundry was dried at a drying place in the garden or on the roof.

The outdoor drying place found a home in the toy world. Old pattern books show drying places of wood; in this century they were made of sheet metal: "Laundry drying place, finest lacquer, can be assembled,

Doll-house washing machines, 1909. (Märklin 4, p. 58)

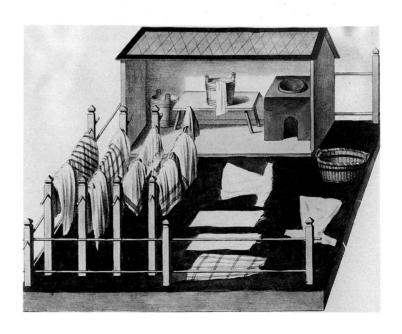

Drying place, 1850-60.
(Nuremberg pattern books, p. 75)

including: 1 sheet-metal base, 1 well, 1 washtub, wringer with rubber rollers, 1 water can, 1 washboard, 1 bench, 4 washpoles, 1 washline, 3 pieces of laundry. 1.80 Marks," the toy dealer Borho of Baden-Baden advertised around 1913. At about the same time, the Viennese toy dealer Anton C. Niessner offered a complete "laundry room with doll, stove, table, wringer, washtub on standard, washing machine and much doll laundry." Sizes are not listed, unfortunately.

Whoever could afford one had a washing machine, even in the doll kitchen. To be sure, washing machines for children were first made relatively late. In the big household, the first ones date from the end of the Eighteenth Century! The early machines were turned by levers or cranks. Similarly working doll-house washing machines were offered by the manufacturers and dealers Herz and Ehrlich in an 1899 "Illustrated Price List" with the following comments:
"Doll-house washing machines, with which one can really wash doll laundry practically, so that these are very suitable for awakening the interest of little girls in household activities".

Doll-house washing machine, 1899. (Herz & Ehrlich, Breslau)

The Märklin Brothers of Göppingen, in 1909, produced "washing machines to be set on the cookstove. A rotating peforated drum in a closed vessel. Practical construction. With wooden handle, 9 or 12 cm high, and also a free-standing washing machine, with a sheet-iron water heater with brass fittings, wooden handles and its own alcohol heater . . . ," and in 1919-21 a "cookstove with laundry kettle".

In the Twenties and Thirties, Böttich-type washing machines with levers and wringer rollers were made of cast zinc for doll kitchens. These machines experienced a noteworthy rebirth after World War II, but soon had to compete with the modern "electric" washing machines. These, of course, had only an old-fashioned vaned wheel that had to be turned by hand, but their external appearance was "genuine": smooth box shape, round window, drainer hose and non-functioning control switch.

Of course there were ironing boards and irons for children, flatirons, coal-heated and finally electric ones too. A child could really iron with them. More symbolic in nature were the small laundry mangles, whether the wooden box type (illustrated by Bestelmeier in 1803

Doll-house washing machine, 1950-60. Sheet metal "Wash-o-mat" with window. Drainer hose and crank. Height 14.5 cm.

and by Niessner about 1910) or the upright type with two rubber rollers, one above the other (shown in, among others, Märklin 1, Nuremberg, 1895; Ullmann & Engelmann, Nuremberg, circa 1900); in toy size they had very little effect.

Laundry mangle, circa 1890. Tin, height 8 cm. (Historical Museum, Frankfurt/Main)

Coal stove with added gas burner,
1927. Amateur photograph,
Regensburg.

The Stove

Cooking on an open fire used great quantities of wood, filled the kitchen with soot and smoke, and cooked the cook with the heat it gave off.

Therefore many attempts were made over the centuries to improve kitchen hearths. Really significant improvements were developed simultaneously around 1800 by economically-minded laymen and scientists, working independently, among them the Hessian pastor Philipp Bus, the cleric and mathematician Josef Danzer in Altötting, and the American Benjamin Thompson, Count Rumford, a physicist and Bavarian statesman in Munich. All of them popularized cooking over a closed firebox. Regulated air supply, concentration of the fire and full use of heat through clever ducting of the smoke were the most important requirements for the construction of a "saving stove".[56]

Characteristic of Danzer's and Rumford's stoves are the depressions, so-called casseroles, that were set into the hearth plate—in one case in a tile, in the other, in an iron plate. In these deep cooking pits one placed the actual cooking pot, so far in that usually only the lid was above the hearth plate.

Along with this type of "casserole stove", Danzer also had the tiled "plate stove" type built. This tiled stove, built of solid masonry from the base of the oven up, enjoyed great popularity among housewives and cooks throughout the whole Nineteenth Century, especially in South Germany, Switzerland and Tyrol, although they had the choice even then of using an iron stove. The development of such stoves made completely of iron, called "cooking machines" in many regions, was based on the discoveries of one J. P. Bérard and the aforementioned Count Rumford, and likewise took place around 1800.

For good heat conducting on both cooking-pit and hearth-plate stoves, pots had to have flat bottoms. But since pots and pans were often rounded, they were finally set over the open fire, even on these stoves. For that purpose, one could remove the right number of concentric rings, according to the size of the pot. The older hanging pots, with their smaller bottoms, hung farther into the fire than the top of the pot projected above. In later hanging pots, used even into the Twentieth Century, the relationship was reversed, with only a small bit hanging into the fire. These last pots also had just a holding ring instead of a built-in step.

Stepped pot, middle of the 19th Century.

Ringed pot to hang in an open fire, light blue, enameled white inside. Height 6 cm.

Family stove (for adults), 1902 (Brockhaus Conversation Lexicon)

Children's Stoves

The simple kitchen hearth for "open fire" is generally to be seen in old pictures of doll kitchens up to the middle of the Nineteenth Century. The sheet-iron stove seems to have found acceptance in children's rooms gradually in the 1830's, and in greater numbers only around the middle of the century. At this time it was already to be found in a variety of designs, for example, that of G. Striebel of Biberach in 1850: a sheet-metal stove with painted-on brick or tile pattern, three brass-painted doors and a baking oven, standing directly on the floor without feet. The Nuremberg pattern books of 1850-60 offer two hearths and one small sheet-metal stove, with a cooking pot standing on the latter's single-door hearth. This kitchen resembles a small sheet-metal kitchen, probably made in Biberich by Rock & Graner around 1850, in the Munich State Museum.[57] The hearth of that one has two openings and two doors on its front. Another small sheet-metal stove of that era, with two doors, baking oven, four cooking pots and round feet, can be seen in another Nuremberg toy catalog from the middle of the Nineteenth Century.

Stoves of this type, handmade of sheet iron, sometimes set with brass, with round feet, other simple types or none at all, with brass doors held closed by a bolt, with baking oven and pots hung deep, usually with high lids, are the old type of doll stoves. They are to be found in the toy sections of many museums. Later, in the last third of the Nineteenth Century, the sheet metal was, for the most part, lavishly stamped, and instead of ordinary feet we find lion's paws of cast or sheet brass.

In their main catalog of 1895, Märklin offered "Cookstoves finely enameled with flower decorations", and shortly after that the Holler'sche Carlshütte firm near Rendsburg and the firm of Herz & Ehrlich in Breslau (1899) also sold "stoves colorfully set off with enamel lacquer".

After 1900 the curved legs and lion's feet were replaced by more or less straight legs, and the stoves, which otherwise still looked quite old-fashioned, stood lower. At this time stoves of every kind were on the market together. Only late in the Twenties and Thirties did smooth white models with undecorated straight legs come to prevail.

Stove with back wall, circa 1900. Sheet metal stamped with a brick-wall pattern, sheet metal utensils. Height (with wall) 16 cm.

Stoves with straight legs.
Left: stove of 1910-20, sheet metal, stamped brick pattern, nickel-plated parts, folding side panels, screwed-on legs. Height (without stovepipe) 26 cm. (Owned by Kim Krier, Frankfurt/Main).

Lower right: stove, 1910-20, sheet metal with bronzed parts, pedestal legs, height (without stovepipe) 12 cm (probably by Moses Kohnstam, Nuremberg).

Stove, circa 1910; sheet metal stamped brick pattern, sheet metal pots, porcelain knobs, height (without stovepipe) 9.5 cm.

Stove, circa 1900; sheet metal, sheet metal pots with brass lids, height (without stovepipe) 8.5 cm.

Doll stove as a photographic prop, circa 1910. Photograph, Ed. Blum/Hoffschild, Frankfurt/Main. (Photo Archives of E. Maas, Frankfurt/Main)

Energy Sources for the Doll Stove

Along with the mass-produced sheet metal stoves, there have also been massive iron stoves for "real heating", until into our century. They were usually made by artisans as journeymen's work, and thus the arrangement of the grate, the ashpit and the baking oven inside, and the exhaust ducts as well, were precise copies of the big stoves.

As long as there was still a flue over the hearth, the children could put their little stove on their mother's big one and make a real fire; the smoke went out through the big flue. Otherwise one could actually cook only outdoors.

Early in their history, the small stoves, which continued to look quite "genuine" from outside, were given an alcohol heating system. In Julie Bimbach's cookbook for the doll kitchen of 1854, a "big cookstove" with a "pretty wine-spirit lamp" is already mentioned.

In most children's cookbooks there are references to the danger of cooking with alcohol. It is no wonder that many memoirs tell of dramatic experiences with this explosive fuel:

"We older children had finally advanced to real cooking after years of cold cooking, and had already proved under observation that we could handle work with alcohol. So no sooner had Berta, our servant girl, gone out on her free Sunday afternoon that we children got ready in the kitchen. Our black inherited sheet metal stove stood on the kitchen table, a mass of dumplings held back from dinner and a piece of roast pork with sauce were brought from the pantry, the cooking pots and doll dishes were gotten out. Once we had filled the long container with alcohol, ignited it and pushed it into the side of the stove, put the salt water on the stove, turned the little dumplings and put the roast pork to warm on the cooler side of the stove,

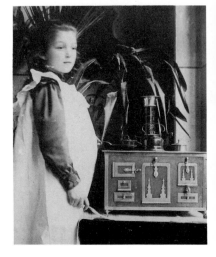

Handmade doll stove, circa 1900. Amateur photo. (Sig. L. Stiegel, Rödermark)

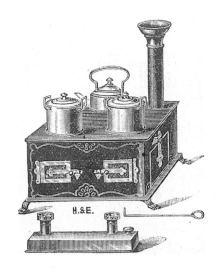

Children's cookstove with spirit lamp including safeguards required by law. (Herz & Ehrlich, Breslau, 1899)

Stove with spirit lamps, circa 1930. (Schminke & Haase, Göttingen)

we stood in a circle around the hearth and stared attentively into the noodle pan. Who wanted to miss the moment when the water finally began to bubble and we could take turns putting one dumpling after another into it. Once we had reached the point where the dumplings all swam in the water, the cooking gradually slowed down, because the amount of alcohol was not enough. Then we had to fill it again. But scarcely had my brother poured some out of the green bottle when there was a bang, and alcohol sprayed out of the presumably too-hot container with loud puffing and made several burning puddles on the floor. Speechless with horror, we stamped out the flames wildly. I still remember today how afraid I was that my checkered brown boots would catch fire.—After the fire was put out we all retired to the bathroom and rubbed our pale faces until our cheeks were red again, so nobody would notice anything. Otherwise we probably would have had to 'cook cold' again!'' (E. P., born 1934)

Alcohol was by no means safe. Finally attempts were even made to connect children's stoves to city gas lines. In 1902 the Märklin Brothers of Göppingen commented on their offering of children's gas stoves as follows: "The way of connecting the gas is similar to that of any gas cookstove. The connecting mouthpiece is located in the end of the so-called protective staff, which here simultaneously forms the ducting to the single valve set on the staff or the burners.''[59]

Cooking with electricity was less dangerous than cooking with alcohol or gas. Because the electro-stoves were expensive—and only put on the market by Märklin in 1909—the alcohol stove could stay on the scene for a long time. The low-priced sheet metal stoves were fueled not only with alcohol but also with ''Teelicht'', and in the last thirty-five years with dry alcohol (''Esbit'') as well.

Stove with baking oven. (Stukenbrok, Einbeck, 1912, p. 131)

Gas stoves for children, left page
Upper left: gas stove (Märklin 1, 1900-1902, p. 147)

Upper right: doll cookstove for gas cooking with regulated burners. (Märklin 4, 1909, p. 96)

Lower left: "gas stove", circa 1910. Sheet metal, lacquered white with light blue trim, for alcohol. Height 13.5 cm.

Lower right: "gas stove", circa 1930. Sheet metal, lacquered white or nickel-plated, for alcohol or "Teelicht", height 14 cm.

Upper center: "gas stove", Thirties. Sheet iron with brass bar, rings, on white lacquered sheet metal table, for alcohol. Height without table, 3.5 cm.

Lower center: gas cooker for children, circa 1913. (Borho, Baden-Baden)

Electro-stoves for children, right page
Above: electro-stove (Märklin 4, 1909, p. 124)

Left center: electro-stove, circa 1920. Enameled white, nickel-plated frame. Height 13 cm.

Right center: doll electric cooker with steam cooking pot, 1930's. Cooker: cast aluminum with wooden handles, height 4.5 cm. Pot: aluminum with bakelite handles. Height 7.5 cm. (Both objects were probably produced in the Fifties.) (Sig. Blömer, Frankfurt/Main)

Lower left: electro-stove, Thirties. Sheet metal, lacquered white, or nickel-plated. With round and oval plates ("Omega" No. 139). Height 17 cm.

Lower right: electro-stove "Bruzelette", circa 1960. Height 13 cm.

Manufacturers of Doll Stoves

Around the middle of the Nineteenth Century, as was noted above, toy manufacturers in Biberach were making sheet metal stoves. Rock & Graner of Biberach was represented at the 1873 Vienna World's Fair, as were the Fürth firms of C. Henglein and Ph. Wüstendörfer. It is specifically recorded that the last two displayed children's cookstoves. For the most part, though, the listings of the firms at industrial and trade exhibitions, and in address books and industry guidebooks, were limited to the collective concept, "toys". Thus it is too that among the many toy manufacturers cited in the guidebook to the Saxon and Thuringian export industry published in Dresden in 1897, only two specifically mention "children's cookstoves and blunt-cornered housewares": the firm of Clemens Kreher in Marienberg, Saxony, and Gustav Fischer & Co. in Zöblitz, in the Erzgebirge; although there certainly were more. Very little is actually known about toy manufacturers.

The companies that manufactured mechanical toys are an exception, and particularly the Märklin Brothers of Göppingen. Their firm's history has been written, and by using the many catalogs that have been reprinted for train collectors, one can also form a picture of their doll stove production at the end of the Nineteenth Century and in the first thirty years of the Twentieth. From the beginning of its existence, that means from 1859 on, the firm made sheet metal housewares for doll kitchens, and may well have been the manufacturer with the overall greatest range of solid doll stoves.

Their stoves are found in dealers' catalogs for both domestic and export trade, as well as in the sales lists of other sheet metal toy manufacturers. For example, the firms of Ullmann and Engelmann and of Moses Kohnstam, both in Nuremberg, listed the better

Stove by F. & R. Fischer, Göppingen, "with brass fittings, guard rail and cut hearth plate. Cooking pots, chocolate cooker, water can of tinplate with brass lids, casserole and kettle of brass." (Fischer price list, Göppingen, 1896) The identical item was offered by the firm of G. R. Schiele, merchants of kitchen and household utensils, Frankfurt. Price list 1884.

Märklin stoves along with simpler products, probably of their own manufacture, in their catalogs from around 1900 to 1928-30.

In this respect, the sheet and cast metal factory of F. & R. Fischer may also be of interest. Like Märklin, it was located in Göppingen. In a catalog of 1896 this firm offered a whole line of doll stoves that differed only minimally from Märklin products. The firm of F. & R. Fischer either worked for the Märklin Brothers or distributed their goods. Or perhaps one of the two firms copied the other.

The most important manufacturers of sheet metal stoves were, among others, the firms of C. W. Engels, Foche/Solingen, Herz & Ehrlich, Breslau, Ernst Plank, Nuremberg, and the Bing Brothers, Nuremberg, and after World War II the firms of Kindler & Briel (Kibri) in Böblingen and particularly that of Fuchs (MFZ) in Zirndorf, near Nuremberg.

Stove by Märklin Brothers, Göppingen. "Sheet iron with cut plate, brass fittings. Guard rail and brass doors. Tinplate cooking utensils with brass lids." (Main catalog, 1895, Märklin 1, p. 99)

Stove by C. W. Engels (Engelswerk), circa 1910. Stamped sheet metal, gold trim, brass fittings, for alcohol, 2 sheet metal pots. Height (without stovepipe) 7 cm. (shown in: Engels, Foche/Solingen, 1913)

Stove by Märklin Brothers, circa 1900. Stamped sheet metal, gold trim, brass fittings, for alcohol, four tinplate pots with brass lids. Height (without stovepipe) 17 cm. (shown in: Märklin 1, p. 147, but with simple lion's feet, and water pot without spigot)

191

Doll Cookbooks

One of the first printed cookbooks for children was probably the "Little Cookbook for the Doll Kitchen or First Instructions in Cooking for Girls from 8 to 14" by Julie Bimbach, published (second edition) in Nuremberg in 1854. According to her own statement, she took most of the recipes from the "Löffler Cookbook", converted them to the right proportions for children, and made a small handwritten cookbook of them, which she put under her own children's Christmas tree with a "big cookstove . . . from your distant grandmother".

Seeing how much her daughters had learned about cooking from this little book (they knew how to make raised dough, butter dough, understood the proportions, and when they grew to the age in which every capable girl must learn to cook, they had already acquired a lot of knowledge that was of good use to them), Julie Bimbach was motivated "to have that little cookbook . . . published for the pleasure and use of many children". A last printing of this cookbook, which had been reprinted over and over, was still on sale in 1947!

In the course of these almost 100 years, one doll cookbook after another was published; most of them also went through many printings, and some were published anew under new titles by other authors and publishers.[58]

A few of the most widespread can be listed here: Christine Charlotte Riedel, "The Little Cook", Lindau, 1854. Marianne Natalie, "Favorite Doll Cookbook for Little Girls", Berlin, 1855. Henriette Löffler, "Löffler's Little Practical Cookbook for the Doll Kitchen", 1860. Marie Schneckenberger, "The Little Cook. Favorite Doll Cookbook for Good Girls", Emil Gutzkow, Stuttgart, no date. Aunt Betty, "Nuremberg Doll Cookbook", Nuremberg, 1892. Anna Jäger,

Title of a children's cookbook of 1891.

Title of a children's cookbook of 1896.

Title of a children's cookbook of 1953.

"Family Daughter's Cooking School for Play and Life", Ravensburg, 1895 (3rd edition). Dolls' and Children's Cookbook, circa 1900, distributed by Märklin. Bertha Heyde, "The Little Doll Cook. Practical Instruction for Cooking", Stuttgart, 1907. Grete Geiringer, "The Doll Cookbook. Twenty-four Foods for the Doll Table", Rikola Publishers, 1922. M. Haarer, "Little Cookbook for Children", Esslingen (1953).

None of these cookbooks, nor any activity book for girls, missed the chance to point out the significance that playtime cooking can have in the development of a young girl into a capable housewife. The suggestions that generally introduced the recipes correspond—in childlike terms—to the introductions of the cooking and housekeeping books for grown women. An example is the story of a capable "Lizzie at the Cookstove": "Lizzie knew exactly what belongs in a completely equipped kitchen . . . and also what is involved in cooking:

"Every little cook must have as big and clean an apron as possible on, and perfectly clean hands. A little cook's hair mustn't fly all over her head either, otherwise some will fall into the soup or the broth, and that ruins the appetite. Cleanliness must be one of the main qualities of little cooks, and therefore she will do better to have one kitchen towel or dustcloth too many in use than too few.

"It goes without saying that all cooking utensils must be very clean, and that containers with edible contents must be protected against dust and flies by covers and cloths is also self-evident. All implements, especially those that are used with anchovies, onions and the like, must be cleaned as soon as possible.

"Keep your kitchen in order, so that you can find the things you need at any moment and don't run around in a sweat and in despair, looking for things while the soup cooks over and the roast burns.

"Punctuality also is a part of cooking, as is precision in following all instructions that are given for preparing a dish. Thrift is also advisable; one should not waste anything needlessly or let anything spoil. Whoever does not learn that in childhood will never learn it."[60]

To the Collector

Collectors expect more than anything else that a book on their field of collecting will provide information that enables them to organize their own objects in a definite time and a definite theme, and if possible, to learn something of their manufacturers and their rarity.

On the subject of toys, it is especially hard to fulfill such hopes, for toys have been made to one and the same pattern over very long periods of time even more than other objects in daily use. Housekeeping toys, in addition, have not represented modern conditions in the world of adults to the same extent as technical toys. One finds utensils and kitchens spread over twenty, thirty and more years. For example, the Viennese toy dealer Anton C. Niessner had a doll kitchen in a

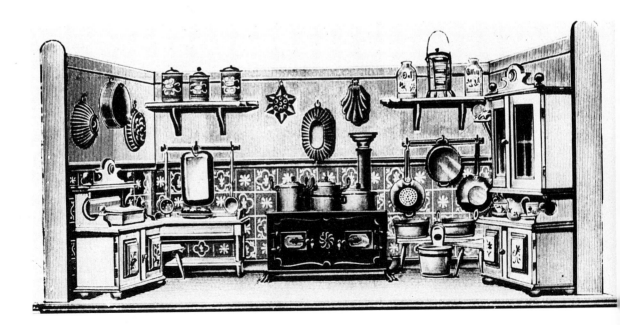

Biedermeier doll kitchen, offered in a catalog of 1893. (Söhlke, Berlin)

Doll kitchen with furnishings in the style of the end of the 19th Century, depicted in toy catalogs of circa 1910 and 1930. (Niessner, Vienna)

housing with tiled floor and furniture in the typical style of the 1890's in his catalog of about 1910. The same doll kitchen appeared unchanged in his 1930 catalog.

One further example out of many is a small doll kitchen in the "Price-Courier" of the "Toy Factory of G. Söhlke, Berlin": This kitchen in Biedermeier style was sold in 1892! If one had enough older catalogs of the Söhlke firm, one could probably trace this kitchen back through many years.

So much for the problem that the catalogs bring up; as far as the originals are concerned, the situation is no less difficult. Doll kitchens come to us very rarely in their original condition. Usually they have been modernized and changed by generations of parents. Every generation has added the mark of its times, at least by removing old objects and adding new ones, but sometimes by painting in new colors, putting on new wallpaper, or even rebuilding the housing. In the doll kitchen shown on page 66, for example, the housing was originally a rather small one in which traces of an old flue can still be seen on the back wall, and the original sink from the time of manufacturing in the 1870's is present. Later the pantry was added on, and after 1900 the bright tile pattern was used to cover the original brownish color; later the kitchen and pantry were equipped with modern furnishings, and later still an electric stove was added.

In the economic depression of the Twenties and Thirties, very much was modernized instead of simply being pushed aside. The drive for practicality and simplicity in this era surely inspired many parents to take the advice, "make the new out of the old", in magazines and instruction books not only in big furniture but in the doll kitchen too, and get rid of "senseless decorations, cloying ornaments and superfluous mouldings." It was probably in this era too that most of the crocheted trimmings were taken off, that ran along the shelves and boards of the kitchen cabinets, in doll kitchens as in big ones, at the end of the Nineteenth Century.

Thus the kitchen that one gets from one's family is not definitely authentic. But it has grown that way and has its own history. Cultural history has nothing static. Every point in time that one picks out is random. It would therefore be most rational to leave a kitchen that has been changed greatly through the generations as it is, instead of striving to regain its "original" condition that cannot be an original condition because present-day conceptions and materials would necessarily be involved.

Of course one sometimes finds doll kitchens that have gone through relatively little change and can be returned to their original conditions without damage. In this case one should be conscientious enough to keep records of how the object was found, what latter-day additions were taken out and which old objects were added. It would be good to document everything photographically, but best of all to mark and preserve everything that belonged to the kitchen.

The same applies to cases in which the collector, overwhelmed by newfound and particularly nice individual items, cannot resist integrating them into a kitchen already at hand. One should mark the new things, just as one does the original ones, and preserve the latter. It is especially regrettable that modern kitchens, somewhat rare in themselves, often are stripped, the housing made to look older and then furnished in correspondingly old style. Perhaps there is something like an obligation to preserve things that are only seldom tangible in the typical character of their times! On the other hand, no collector needs to be too ashamed of mistakes once made. He is not alone, for almost all of us have already committed "sins" once. (For example, I remember all kinds of dispassionate changes in doll kitchens that I once made for a children's book!) Besides, even museum people are human. I know of one museum doll kitchen, among others, that was furnished with lovely sheet metal utensils at an exhibition in 1961. In a 1975 book it appeared with copper, brass and especially with light blue enamel utensils that made much more of an "impression".

Now for another group of doll kitchens that the collector encounters in ever-greater quantities: kitchens that have passed through collectors' and/or dealers' hands. Only in exceptional cases can one presume an original condition. In general, one will have to proceed on the basis of experience and assume that everything has been "cobbled up". It is not uncommon that one

Doll stove, circa 1950. Sheet metal printed blue and white. Style of the Twenties, height (with back wall) 17.5 cm. Keim Co., Nuremberg.

196

even obtains an empty housing that one will furnish little by little. In these cases one naturally is not responsible for "irreplaceable cultural treasures". The best examples for a "correct" furnishing would be catalog pictures from the time of the housing or the furnishings. For if one likes to collect systematically and derives pleasure from making something complete and relatively "in tune", then surely it is not right to say, "Doll kitchens have grown through generations, and therefore one can add everything at random." But if the collector has no claim to being systematic, he can add anything that pleases him in such a case. Then he does not need any scientific alibis and the like either. It is simply his right to play with the things in a creative way and enjoy his possessions. As everywhere in life, there are many courses, of which neither the one nor the other must be bad.

Protection and Restoration

For whatever reasons and in whatever manner the collector may be involved with toys, he is confronted with the problems of protection and restoration, and in a certain sense even "obligated" to care and, if possible, to preserve something historic. First of all, two principal matters must be differentiated, on the one hand, protection, and on the other, restoration.

Protection is actually a matter of preserving an object and safeguarding it from further decay. Since toys consist of the most varied materials, and since every old object needs individual treatment, no detailed recipes can be given in this brief discussion. We can only say something basic: along with the object—as far as possible—its history should also be preserved. Everything that can be learned about a toy from its previous owner should be noted down: to which child the toy belonged? (sex, year of birth, conditions of life, that is, the social position of the parents, number of siblings, place of residence). To whom did the toy belong before that? What else can the previous owner tell?

Newly acquired toys are to be examined for damage. If one finds traces of moths, silverfish, fur or cabinet beetles or woodworms, one should expose the objects to the insecticide of a generous portion of mothballs for several weeks in a tightly sealed plastic bag. In doing this, one can prevent direct contact with the toy by putting acid-free silk paper in between. For long-term storage, though, plastic is absolutely not to be used, as a kind of "softener" is gradually released from the material, and this can really damage old material, especially textiles and paper.

Every toy is dusted first, no matter what material it is made of. The dust can be removed with a soft brush, or perhaps even with a vacuum cleaner (with weak suction!). In the latter case, gauze or some other fine, loose weave must be put over the object, so that loose pieces do not disappear into the vacuum.

Cleaning with water or other solvents is often necessary, for dirt can also be destructive under some conditions. But general rules cannot be made for high-risk cleaning.

Patching or mending with needle and thread, though sometimes unavoidable, also usually requires consultation with specialists.

In principle, one should use only means that one can, if necessary, reverse. For example, never use "super glue" or self-adhesive materials (such as plastic tape), since these leave ineradicable marks after a time.

Restoration means recreating the original condition or something approaching it. In many cases, the original condition is no longer known or no longer attainable for technical reasons. In such cases one can only achieve a presumed or newly made "original condition". Even under favorable conditions it should be considered whether a return to an earlier state of a toy is really right.

Then too, the development that an object has experienced is even more interesting.

All in all, one should have courage with a worn old toy. A toy does not need to be perfect, for it is not precious in the same sense as old glassware or porcelain figures. Its value is not ruined by a crack, for its value lies on another plane: if one thinks in terms of cultural history, a toy is interesting because it is capable of telling something; if one thinks as a dedicated collector, then it is lovable when it shows signs of children and reveals something of their play.[61]

Anonymous photograph, circa
1925.

Footnotes

1. Marie Nathusius, Lebensbild der heimgegangenen, Vol. I, Mädchenzeit. Halle, 1867, p. 19.
2. Georg Ebers, Die Geschichte meines Lebens. Stuttgart, 1893, p. 120.
3. Ludwig Ganghofer, Buch der Kindheit. Stuttgart, 1922, p. 78.
4. Anna Jäger, Haustöchterchens Kochschule. Ravensburg, 1896, p. 28ff.
5. Heidi Müller, Dienstbare Geister. Berlin, 1981, p. 31ff.
6. Michael Andritzky et al., Lernbereich Wohnen. Reinbek, 1979, p. 259.
7. Such a house is known as a "smokehouse"; the tar protects the beams from insects. See also: Alfred Faber, 1000 Jahre Werdegang von Herd und Ofen. Munich, 1950, p. 5ff.
8. Two doll kitchens can be cited as examples, one owned by the Schmarje family, in the Altona Museum in Hamburg (illustrated in H. Schwindrazheim, Altes Spielzeug aus Schleswig-Holstein, Heide, 1957, #27), the other from the von Leonhardi family of Grosskarben (illustrated in Chr. v. d. Marwitz, Spielzeug aus Frankfurter Familienbesitz, Frankfurt, 1965, p. 109 & title).
9. J. Rottenhöfer, Die gute bürgerliche Küche. Munich, (1863), p. 4 (1st Ed., 1858).
10. Hermann Kaiser, Herdfeuer und Herdgerät im Rauchhaus. Wohnen damals. Cloppenburg, 1980, p. 54.
11. Marie Leske, Beschäftigungsbuch für Mädchen. Leipzig, 1865, p. 10 & p. 105.
12. Carmen Sylva, quoted by Paul Hildebrandt, p. 118.
13. Princess Marie zu Erbach-Schönberg, Entscheidende Jahre. Darmstadt, 1923, p. 66ff.
14. Marie Leske, 1865, p. 108.
15. Chr. v. d. Marwitz, Der kleinen Kinder Zeitvertreib. Darmstadt, 1967, p. 97.
16. Elly Gregor, Joh. von Sydow, Lieschens Puppenstube. Leipzig, (1894), p. 40ff (1st Ed. 1888).
17. Josef August Lux, Die moderne Wohnung und ihre Ausstattung. Vienna, 1905, p. 50ff.
18. Rudolf Mehringer, Das deutsche Haus und sein Hausrat. Leipzig, 1906, p. 97.
19. Paul Hildebrandt, Das Spielzeug im Leben des Kindes. Berlin, 1904, p. 118.
20. Franz Schuster, Eine eingerichtete Kleinstwohnung. Frankfurt/Main, 1927.
21. Catherine E. Beecher & Harriet Beecher Stowe, The American Woman's Home. New York, 1869, p. 33, quoted by Sigfried Gideion, Die Herrschaft der Mechanisierung, Frankfurt, 1982, p. 563.
22. Kornelia Kopp, Ein Mädel will heiraten. Leipzig, (1938), p. 6ff.
23. ibid, p. 4ff.
24. Schönheit des Wohnens, reprinted from: Der soziale Wohnungsbau in Deutschland. Vol. 2-41, Berlin, (1941), p. 1ff.
25. Das Spielzeug, April-May 1943, p. 6.
26. Sigfried Gideion, Die Herrschaft der Mechanisierung. Frankfurt, 1982, p. 651.
27. Amtlicher Bericht über die Allgemeine Deutsche Gewerbe-Ausstellung zu Berlin, 1844, vol. III. Berlin, 1846, No. 2960.
28. Preis-Courant von Götzinger, Merseburg, 1843. (Thanks to Christa Pieske, Lübeck)
29. Gertrud Meyer, Das Spielwarenindustrie im Erzgebirge. Leipzig, 1911, p. 34ff.
30. Illustration in Die Kunst, 13 (1910).
31. Rundschau über Spielwaren, Galanteriewaren und Sportartikel 161 (1914), p. 2646 (Thanks to Werner Schröder, Düsseldorf)

32. Striebel, Biberach um 1850, p. 103 (City Archives, Biberach a. R.)

33. Luise Wilhelmi & William Löbe, Illustrirtes Haushaltungs-Lexicon. Strassburg, 1884, p. 401.

34. Goethe's Works, Vol. 8, Dichtung und Wahrheit. Leipzig, 1903, p. 8ff.

35. Bogumil Goltz, Buch der Kindheit. Berlin, 1854, p. 250.

36. Wolfgang Gatzka, WHW-Abzeichen. Munich, 1981, p. 97.

37. Illustrirte Frauenzeitung, June 16, 1883, p. 223.

38. Manfred Bachmann, Holzspielzeug aus dem Erzgebirge. Dresden, 1984, p. 195.

39. Offizieller Ausstellungsbericht, Vol. 3. Vienna, 1873, p. 22.

40. Paul Hildebrandt, Das Spielzeug im Leben des Kindes. Berlin, 1904, p. 117.

41. Wathner's Eisenund Eisenwaren-Kenner. Graz, 1885, p. 189.

42. Bestelmeier, Nuremberg, 1803, p. 7.

43. Amtlicher Bericht über die Allgemeine Deutsche Gewerbe-Ausstellung zu Berlin, 1844. Berlin, 1846, Vol. II, p. 385ff.

44. ibid, Vol. II, p. 258. In 1844 Justus Assmann "provided 34 objects of tinplated sheet iron for exhibition".

45. Götzinger, Merseburg, 1843.

46. Deutsche Gewerbe-Ausstellung Berlin 1844. Berlin, 1846, Vol. II, p. 385.

47. ibid, Vol. III, p. 119.

48. A number of Märklin catalogs were reproduced by C. Baecker, D. Haas, C. Jeanmaire. In the series "Die anderen Nürnberger" edited by C. Baecker & D. Haas, girls' toys are likewise included in some volumes.

49. Wathner, Graz, 1885, p. 200.

50. Deutsche Allgemeine Gewerbe-Ausstellung Berlin 1844. Berlin, 1846, Vol. II, p. 122. See also Brigitte ten Kate, Email. Weil der Stadt, 1983, p. 43 & p. 83.

51. Wathner, Graz, 1885, p. 193ff.

52. Brigitte ten Kate, 1983, p. 85. The Baumann Brothers put a draw press into operation in 1879.

53. "And finally we are rightfully proud of our beautiful cooking utensils, a delicate light blue on the outside, pure white enamel inside . . ." Quoted by Anna Jäger, Haustöchterchens Kochschule. Ravensburg, 1896, p. 193.

54. Brigitte ten Kate, 1983, p. 178.

55. Märklin 4, pp. 93-97.

56. Alfred Faber, Entwicklungsstufen der häuslichen Heizung. Oldenburg, 1957, p. 141ff. The development of the saving stove is not attributable to Count Rumford alone.

57. Small sheet metal kitchens, circa 1850, illustrated in: C. Baecker et al., Vergessenes Blechspielzeug. Frankfurt/Main, 1982, p. 54.

58. For details of the history of publication by Werner Schröder, see the Bibliography.

59. Märklin 1, p. 150.

60. Elly Gregor et al., Lieschens Puppenstube, 1894, p. 44ff.

61. This chapter is taken from the exhibition catalog: A. Junker, E. Stille, Spielen und Lernen. Spielzeug ud Kinderleben in Frankfurt 1750-1930, Frankfurt/Main, 1984.

Bibliography

In addition to the literature listed here, cookbooks and housekeeping books, catalogs of housewares, photo albums and publications on interior architecture have been referred to.

Adressbuch der gesamten sächisch-thüringischen Industrie. Dresden, 1901.

Addressbuch Deutscher Exportfirmen. Berlin/Leipzig, 1885.

Andritzky, Michael, & Gert Selle (Ed.), Lernbereich Wohnen, Vol. I, Reinbek bei Hamburg, 1979.

Bauer, Ingolf, Hafnergeschirr, Katalog des Bayerischen Nationalmuseums (6). Munich 1980.

Baemerth, Karl, Bibliographie zur hessischen Keramik. Otzberg, 1980.

Beecher, Catherine E. & Harriet Beecher Stowe, The American Woman's Home. New York, 1869.

Benker, Gertrud, Altes bäuerliches Holzgerät. Munich, 1976.

Benker, Gertrud, Küchengeschirr und Essensbräuch. Regensburg, 1977.

Brödner, Erika, Moderne Küchen. Munich (1950).

Brödner, Erika & Rudolf Schlick, Heimgestaltung. Darmstadt (1959).

Deutsche Gewerbe-Ausstellung (Amtlich Bericht über die) zu Berlin im Jahre 1844. Berlin, 1846.

Elcho, Rudolf, Ein Gang durch die Hygiene-Ausstellung zu Berlin. In: Illustrirte Frauenzeitung, July 16, 1983, p. 222ff.

Faber, Alfred, 1000 Jahre Werdegang von Herd und Ofen. Munich, 1950.

Faber, Alfred, Entwicklungsstufen der häuslichen Heizung. Oldenburg, 1957.

Flink, Maria, Die perfekte Köchin. Ein Kochbuch einfach, deutlich und bewährt. Dillenburg, 1854.

Frauenalltag und Frauenbewegung in Frankfurt 1890-1980. Info-Blätter zur Ausstellung. Historisches Museum, Frankfurt/Main, 1981.

Führer durch die Sächsisch-Thüringische Export-Industrie. Dresden, 1897.

Gatzka, Wolfgang, WHW-Abzeichen. Munich, 1981.

Gehren, Wilhelmine von, Küche und Keller. Ein hauswirtschaftliches Nachschlagebuch, zugleich ein Ratgeber. Berlin (1904).

Gideion, Sigfried, Die Herrschaft der Mechanisierung. Frankfurt/Main, 1982.

Grasmann, Lanbert, Kröninger Hafnerei. Regensburg, 1978.

Grein, Gerd J., Materialien zur Töpferei in Hessen I, Keramik aus Marjoss im Spessart. Otzberg, 1981.

Grein, Gerd J., Materialien zur Töpferei in Hessen II, der Kunsttöpfer Valentin Braun aus Urberach. Otzberg, 1983.

Kaiser, Hermann, Herdfeuer und Herdgerät im Rauchhaus. Wohnen damals. Museumsdorf Cloppenburg, 1980.

Kate-von Eicken, Brigitte ten, Email für Haushalt und Küche. Weil der Stadt, 1983.

Kate-von Eicken, Brigitte ten, Küchengeräte um 1900. Weil der Stadt, 1979.

Koch, Alexander, Das schöne Heim. Darmstadt, 1920.

Kopp, Kornelia, Ein Mädel will heiraten. Ratschläge für die Aussteuer. Leipzig (1938).

Lehmann, Gustav, Die Einrichtung der Bürgerlichen Wohnung. Munich, 1924.

Lux, Josef August, Die moderne Wohnung und ihre Ausstattung, Vienna, 1905.

Mehringer, Rudolf, Das deutsche Haus und sein Hausrat. Leipzig, 1906.

Meyer, Gertrud, Die Spielwarenindustrie im Erzgebirge. Leipzig, 1911.

Müller, Heidi, Leben und Arbeitswelt städtischer Dienstboten. Berlin, 1981.

Müller-Wulckow, Walter, Die deutsche Wohnung der Gegenwart. Königstein/Taunus, 1930.

Naumann, Joachim (Ed.), Hessische Töpferei zwischen Spessart, Rhön und Vogelsberg, Katalog der Staatlichen Sammlungen Kassel. Melsungen (1975).

Neues Bauen, Neues Gestalten. Das neue Frankfurt, die neue Stadt, eine Zeitschrift zwischen 1926 und 1933. VEB Verlag der Kunst, Dresden, 1984.

Neundörfer, Ludwig, Wie wohnen? Königstein/Taunus, no date. Offizieller Ausstellungsbericht. Weltausstellung Wien, Vol. 3. Vienna, 1873.

Petzold, K. Figala, Sir Benjamin Thompson, Graf von Rumford (1753-1814). In: Kultur und Technik, December 1983, p. 235ff.

Pröpper, Ludovica von, Eigner Herd. Guter Rath für junge Hausfrauen und solche, die es werden wollen. Frankfurt/Main (1896).

Rottenhöfer, J., Die gute bürgerliche Küche in allen ihren Theilen. Munich (1963); 1st Edition, 1858.

Rumford, Benjamin Graf von, Ueber Küchen-Feuerherde und Küchengeräthe, (Kleine Schriften politischen, ökonomischen und philosophischen Inhalts, Vol. 3) Weimar, 1803.

Schönheit des Wohnens, Deutscher Hausrat mit dem Gütezeichen der DAF. Reprinted from: Der soziale Wohnungsbau in Deutschland, 2/41. Berlin, 1941.

Schuster, Franz, Eine eingerichtete Kleinstwohnung. Frankfurt/Main, 1927.

Späth, Kristine, Töpferei in Schlesien, Bunzlau und Umgebung. munich, 1979.

Stille, Eva & Peter Beitlich, Aus der Küche um 1900. Munich, 1978.

Stille, Eva, Trautes Heim Glück allein. Gestickte Sprüche für Haus und Küche. Frankfurt/Main, 1979.

Tiedemann, Lotte, Küche und Hausarbeitsraum. Frankfurt/Main (1963).

Wahrlich, Hermann, Wohnung und Hausrat. Munich, 1908.

Wathner's praktischer Eisenund Eisenwaaren-Kenner oder gründliche Anleitung zur Kenntnis der Eisenwaaren und deren Gattungen. Ed. Josef Tosch. Graz, 1885.

Weltausstellung 1873 Wien. Offizieller General-Katalog. Wien, 1873.

Wiessner, Paul, Die Anfänge der Nürnberger Fabrikindustrie. Frankfurt/Main, 1929.

Wilhelmi, Luise & William Löbe, Illustrirtes Haushaltungs-Lexicon. Strassburg, 1884.

Wiswe, Mechthild, Hausrat aus Kupfer und Messing. Munich, 1979.

Children, Childhood, Toys

Ariès, Philippe, Geschichte der Kindheit. Munich, 1975.

Baecker, Carlernst & Christian Väterlein, Vergessenes Blechspielzeug. Frankfurt/Main, 1982.

Bayer, Lydia, Das Spielzeugmuseum der Stadt Nürnberg. Nuremberg, 1978.

Betty, Tante, Nürnberger Puppenkochbuch. Nuremberg, 1909.

Bimbach, Julie, Kochbüchlein für die Puppenküche oder erste Anweisung zum Kochen für Mädchen von 8-14 Jahren. Nuremberg, 1858.

Davidis, Henriette, Puppenköchin Anna. Praktisches Kochbuch für kleine und grosse Mädchen. Adapted by Emma Heine. Leipzig (1891).

Ebers, Georg, Die Geschichte meines Lebens. Stuttgart, 1893.

Erbach-Schönberg, Fürstin Marie zu, Entscheidende Jahre. Aus meiner Kindheit. Darmstadt, 1923.

Fraser, Antonia, Spielzeug. Oldenburg/Hamburg, 1966.

Fritzsch, Karl Ewald & Manfred Bachmann, Deutsches Spielzeug, Leipzig, 1965.

Ganghofer, Ludwig, Buch der Kindheit. Stuttgart, 1922.

Goethes Werke, Vol. 8, Dichtung und Wahrheit. Ed. A. Stern. Leipzig, 1903.

Goltz, Bogumil, Buch der Kindheit. Berlin, 1854.

Gregor, Elly & Johanna von Sydow, Lieschens Puppenstube. Leipzig (1884).

Haarer, M., Kleines Kochbuch für Kinder. Esslingen (1953).

Hildebrandt, Paul, Das Spielzeug im Leben des Kindes. Berlin, 1904. New edition, Berlin, 1979.

His, H. P., Altes Spielzeug aus Basel. Basel, 1979.

Jäger, Anna, Haustöchterchens Kochschule für Spiel und Leben. Ravensburg (1896).

Kaut, Hubert, Alt-Wiener Spielzeugschachtel. Vienna, 1961.

Kinderspielzeug. Katalog des Museums für Völkerkunde und des Museums für Volkskunde. Basel, 1964.

King, Constance Eileen, Das grosse Buch vom Spielzeug. Zollikon, 1978.

Kloos, Werner, Bremer Kinder und ihr Spielzeug. Bremen, 1969.

Korff, Gottfried, Puppentheater als Spiegel bürgerlicher Wohnkultur. In: Wohnen im Wandel. Wuppertal, 1979.

Kutschera, Volker, Spielzeug. Spiegelbild der Kulturgeschichte. Salzburg, 1975.

Leske, Marie, Illustrirtes Spielbuch für Mädchen. Leipzig, 1865.

Marwitz, Christa von der, Der kleinen Kinder Zeitvertreib. Darmstadt, 1967.

Marwitz, Christa von der, Spielzeug aus Frankfurter Familienbesitz. Frankfurt/Main, 1965.

Aus Münchner Kinderstuben 1750-1930. Ausstellungskatalog Münchner Stadtmuseum. Munich, 1976.

Schneckenburger, Marie, Die kleine Köchin. Allerliebstes Puppen-Kochbüchlein für kleine brave Mädchen. Stuttgart, no date.

Schneider, Jenny, Spielzeug des 18. und 19. Jahrhunderts. Katalog des Schweizerischen Landesmuseums Bern. Bern, 1969.

Schröder, Werner, Historischer Abriss über die Geschichte der Puppen-Kinder-Kochbücher. In: Börsenblatt 78 (1981), p. 2257ff.

Schwindrazheim, Hildemarie, Altes Spielzeug aus Schleswig-Holstein. Heide in Holstein, 1957.

Senft, Otto, Die Metallspielwarenindustrie und der Spielwarenhandel von Nürnberg und Fürth. Erlangen, 1901.

Die deutsche Spielwarenindustrie. Herausgegeben vom Ausschuss zur Untersuchung der Erzeugungsund Absatzbedingungen der deutschen Wirtschaft. Berlin, 1930.

Thumm-Soffel, Britta, Puppengeschirr—Zeitgeschichte in der Puppenküche. In: Puppen & Spielzeug, Vol. 4, No. 3 & 4 (1979), Vol. 5, No. 1, 3 & 4 (1980).

Tintner, Erwin, Das Puppenkochbuch, 24 Speisen für den Puppentisch. Vienna, Leipzig, Munich, no date.

White, Gwen, Toys, Dolls, Automata. Marks and Labels. London, 1975.

Wich, J. P., Steckenpferd und Puppe. Nördlingen, 1847.

Wilckens, Leonie von, Tageslauf im Puppenhaus, bürgerliches Leben vor dreihundert Jahren. Munich, 1956.

Wilckens, Leonie von, Das Puppenhaus. Vom Spiegelbild des bürgerlichen Hausstands zum Spielzeug für Kinder. Munich, 1978.

List of the most important toy pattern books and catalogs

Bestelmeier, Georg Hieronimus, Magazin. Nuremberg, 1803. Zürich, 1979 (Reprint).

Au Bon Marché, Etrennes-Jouets, Paris, 1913. In: Toys, Dolls, Games, Paris, 1903-1914, London, 1981 (Reprint).

Borho, H., Katalog über Spielwaren, Baden-Baden, circa 1913.

Engels, C. W., Engelswerk Stahlwaren-Fabrik, Weihnachtspreisliste. Foche/Solingen, 1913.

Faber, Arthur, Holzwarenfabrik, Katalog Küchen/Haushaltartikel/Möbel. Bietigheim/Württemberg, 1910.

Fischer, F. & R., Blechund Metallwaren-Fabrik, Illustrierte Preisliste. Göppingen, 1896.

Gamage's, Catalogue, London, 1902-1906. In Mr. Gamage's Great Toy Bazaar. London, 1892 (Reprint).

Götzinger, August, PreisCourant von Spielsachen eigner Fabrik. Merseburg, February 1843.

Herz & Ehrlich, Fabrikation und Export, Illustrirte Preisliste No. 6. Breslau (1899).

Hesmer & Möllhoff, Aluminium-Spielwaren. Bärenstein/Westphalia, 1927.

Holzspielsachen-Musterbuch, Nuremberg, circa 1870. (Owned by the Nuremberg City Library).

Kohnstam, Moses, Katalog, Nuremberg, 1928-1930. In: Carlernst Baecker, Dieter Haas, Die Anderen Nürnberger, Technisches Spielzeug aus der "Guten Alten Zeit", Vol. 5. Frankfurt, 1976.

Lutz, Ludwig, Auszüge aus Musterbüchern, Ellwangen, 1846-91. In: C. Baecker, C. Väterlein, vergessenes Blechspielzeug. Frankfurt/Main, 1982.

Märklin, Gebrüder, Hauptkatalog, Abt. I, Einrichtungs-Gegenstände für Kinderküchen etc. Kochherde und Küchen. Göppingen, 1895, 1900-1902. In: Carlernst Baecker, D. Haas, C. Jeanmaire, Märklin 1, Technisches Spielzeug im Wandel der Zeit. Frankfurt, 1975.

Märklin, Gebrüder & Co., Katalog M9, Göppingen, 1909-1912. In: C. Baecker, D. Haas, C. Jeanmaire, Märklin 4, Technisches Spielzeug im Wandel der Zeit. Frankfurt/Main, 1978.

Märklin, Gebrüder, Hauptkatalog (Spiele), Göppingen 1919-1921. In: C. Baecker, D. Haas, C. Jeanmaire, Märklin 6, Technisches Spielzeug im Wandel der Zeit. Framkfurt/Main, 1980.

Meyer, E. L., Auswahl: Liste 24. Hildesheim, 1925.

Meyer, E. L., Auswahl: Liste 26S, Spielwaren. Hildesheim (1927).

Meyer, E. L., Auswahl KG: Spielwaren-Preisliste. Hildesheim, 1968.

Müller, Wilhelm Aug., Puppen, Spielwaren, Christbaumschmuck. Sonneberg/Thuringia, circa 1925.

Niessner, Anton C., Spielwaren-Katalog. Vienna, circa 1910.

Niessner, Anton C., Alle Kinder Wünschen, Vienna (1930).

Nürnberger Musterbücher um 1850/60. In: Christa Pieske, Schönes Spielzeug aus alten Nürnberger Musterbüchern. Munich, 1979 (Reprint).

Nürnberger Spielzeugkatalog, mitte 19. Jahrhundert. Owned by Det Danske Kunst-Industriemuseum. Copenhagen.

Printemps, Grands Magazins du, Leksaker. Paris, 1891-92, printed for Scandinavia. Reprint of the Toy Museum in Stockholm, 1980.

Samaritaine, La, Jouets, Paris, 1906, 1909, 1910, 1913, 1914. In: Toys, Dolls, Games. Paris, 1903-1914. London, 1981 (Reprint).

Schminke & Haase, Grosshandlung, Katalog. Göttingen, circa 1930.

Söhlke, G. Nachfolger Spielwaren-Fabrik, Illustrirter Katalog und Preis-Courant. Berlin (1893).

Sonneberger Spielzeugmusterbuch (Spielwaren-Mustercharte von Johann Simon Lindner, Sonneberg, 1831). Leipzig, 1979 (Reprint).

Spielzeuge, 700, Nuremberg, circa 1954. Düsseldorf 1979 (Reprint).

Spielzeuge, 1200, Nuremberg, circa 1954. Düsseldorf, 1979 (Reprint).

Striebel, Gottfried, Musterbuch, Biberach an der Riss, circa 1850. In: C. Baecker, C. Väterlein, Vergessenes Blechspielzeug. Frankfurt/Main, 1982, and City Archives, Biberach an der Riss.

Stukenbrok, August, Illustrierter Hauptkatalog. Einbeck, 1912. Reprint of the original in the Einbeck City Archives, Olms Presse, Hildesheim, 1973.

Stukenbrok, August, illustrierter Hauptkatalog. Einbeck, 1926. Reprint of the original in the Einbeck City Archives, Olms Presse, Hildesheim, 1974.

Aux Trois Quartiers, Jouets, Paris, 1906. In: Toys, Dolls, Games, Paris, 1903-1914. London, 1981 (Reprint).

Ullmann & Engelmann, Preis-Courant. Fürth/Bavaria, circa 1900. In: C. Baecker, D. Haas, C, Väterlein, Die Anderen Nürnberger, Technisches Spielzeug aus der "Guten Alten Zeit", Vol. 6, Frankfurt/Main,1981.

Ville de St. Denis, Grands Magasins de la, Jouets, Paris, 1903, 1910, 1911, 1914. In: Toys, Dolls, Games, Paris, 1903-1914. London, 1981 (Reprint).

Waldkirchner Spielzeugmusterbuch, Das, (Der Spielwarenverlag Carl Heinrich Oehme, Waldkirchen, circa 1850). Leipzig, 1977 (Reprint).

A. Wertheim, Warenhaus, Berlin, Mode-Katalog 1903/04. Reprint: Olms Presse, Hildesheim, 1982.

In addition, catalogs of the last thirty years from various stores, mail-order and toy firms have been used.

Photographs

All illustrated objects came from the collection of E. Stille of Frankfurt and were photographed by Severin Stille, Frankfurt, unless it is otherwise noted in the caption.

Stove, circa 1900. Stamped sheet
iron, brass doors and lion's feet.
For alcohol; tinplate cooking
utensils. Height (without
stovepipe) 12 cm.

Our Bairns

Glimpses of Tyneside's Children
c.1850 – 1950

Joan Foster

Published by Newcastle Libraries & Information Service

Acknowledgements

I am most grateful to Angela Forster of the Local Studies Section, Newcastle City Library, for allowing me to use her idea for this book, to the other members of staff for their support and help and also to Anna Flowers of the Publications Section. David, my husband, has been as encouraging as ever. Anne Gladders has patiently typed the script. My daughter, Clair, diligently proof read the text.

The following organisations and groups have provided me with detailed and careful information. I appreciate their help and interest: Ashfield Nursery, Barnardo's North East, Central Newcastle High School, Children North East, the NSPCC, the Royal Victoria Infirmary, Tyne & Wear Archives Service and West Newcastle Local Studies Group.

I have depended considerably on oral evidence and would like to thank: Ian Atkinson, Grace Davidson, Jimmy Forsyth, Jacqueline Gibson, Hilda Hope, Dr R.H. Jackson, Maureen McDade, Norman Pallace, James Perry, Terry Quinn and Desmond Walton. Hedley Chambers generously allowed me to use the unpublished memories of his mother, Doris Chambers.

Most interesting letters and books on the sad history of child migrants have arrived from Australia from: Belinda Barry, Gordon Grant, Ruby Middling, Richard Stewart and Lionel Welsh.

Many people have provided me with leads, clues and contacts. I am indebted to them all but due to space can name but a few: Anne Craft, Professor Brian Hackett, Tom Hadaway, Dorothy Hind, Mike Kirkup, Sue Lewis, Frank Manders, the late Dr F.J.W. Miller, Sally Odd, Lady Ridley.

I have really enjoyed researching and writing this book, but I am conscious of many areas that I have been unable to examine and explore. The subject is wide and it has been necessary to be selective. Any errors in the text are my responsibility.

Photographic acknowledgements

Front cover: Todd's Nook School, 1897.

Back cover: children playing in the back-lane, Ancrum Street, Spital Tongues, c.1920s.

All the illustrations come from the Local Studies Collection, Newcastle City Library except for the following: Barnardo's North East, page 44; Leonard Bell, back cover and page 26; Children North East, page 12 ; Jacqueline Gibson, page 41; Mrs R. Hunter, page 30; Newcastle Chronicle & Journal Ltd., page 39; Geoff Phillips, page 13; Terry Quinn, pages 22, 42; Royal Victoria Infirmary, page 47; Tyne and Wear Archives, pages 37, 45; Desmond Walton, page 25; West Newcastle Local Studies, front cover and pages 23, 24, 27, 28, 31, 32, 36.

Extracts and quotes from printed books and newspapers remain copyright of the sources cited.

Cover design by A.V. Flowers

For Thomas William Lawrance, born 16 January 1995

ISBN 1 85795 023 2

© Joan Foster, 1997
City of Newcastle upon Tyne,
Community & Leisure Services Department,
Newcastle Libraries & Information Service, 1997

Cataloguing-in-Publication Data: A catalogue record for this book is available from the British Library.

Printed by Bailes the Printer, Houghton-le-Spring

Contents

Photographs

Children in Sandgate, c.1898. Even the youngest of children had work to do. A small child is carrying firewood, either to heat her own home or to sell. An older girl is keeping an eye on the little ones. The bare feet and poor clothing indicate their hard lives, and the open drain is evidence of the insanitary living conditions.

Introduction

The photographic collection of Newcastle Libraries and Information Service, Local Studies Section contains a number of evocative pictures of Newcastle's children. Where there is little or no written evidence the early lantern slides and later photographs show images of childhood from the mid-19th century to the early 1950s. The lens of the camera recorded moments of childhood that would otherwise have disappeared. The first photographs in this book show the legacy of Victorian 'Hard Times' for many of the city's barefoot and ragged children, living and playing on the streets.

The hearts and consciences of local businessmen and concerned individuals were roused and holidays, breakfasts and shelters were funded. From these roots developed local charitable ventures which are still working today for the care of children in the area.

School, church and home were the cornerstones of most children's lives, and these are illustrated by photographs and written accounts.

The two World Wars brought disruption and changes. Many of Newcastle's mothers provided the labour force for the city's armaments and engineering works and nursery care was gradually introduced for the youngest children. For the older ones schooldays were interrupted and during World War II most experienced the profound effects of evacuation.

The book concludes with the significant advances in Newcastle's child health care. From a position of having one of the highest national levels of death from tuberculosis and worst average for infant mortality, Newcastle came to lead the way in the treatment of children in hospital and child welfare clinics.

The Industrial Revolution

During the 19th century, the brutalisation that accompanied the early years of the Industrial Revolution swung the pendulum to the disadvantage of children, particularly those of the working classes. Here was an expendable, unprotected and cheap labour force for the textile mills, potteries and mines. The children of Newcastle upon Tyne and its environs were amongst the many who suffered.

Apart from the appalling working conditions, the children of the labouring poor inevitably lived in cramped and distressing slum housing. *The Report on the Sanitary Conditions of Newcastle, Gateshead, North Shields, Sunderland, Durham and Carlisle* (1845) by Dr D.B. Reid is graphic. Newcastle's 'fever districts', the Quayside and Sandgate areas, are described vividly:

> The streets most densely populated by the humbler classes are a mass of filth where the direct rays of the sun never reach. … To take a single example of one of the more extreme cases shown to me when visiting them during the day, a room was noticed with scarcely any furniture and in which there were two children of two and three years of age absolutely naked, except for a little straw to protect them from the cold, and in which they could not have been discovered in the darkness if they had not been heard to cry.

Such conditions led nationally and locally to an exceptionally high mortality rate among children. From every 1,000 children born in England and Wales in 1851 only 522 ever reached the age of five. Against this background, moral guidance and parental upbringing were linked to the expectation of early death. The influential evangelical movement of the period focused on the prospect of either heaven or hell as the determining factor in the training of young children. Susanna, the mother of Charles and John Wesley, wrote to John as follows:

> Let a child, from a year old, be taught to fear the rod and to cry softly. In order to do this, let him have nothing he cries for; absolutely nothing, great or small; else you undo your own work. At all events, from that age, make him do as he is bid, if you whip him ten times running to effect it. Let none persuade you it is cruelty to do this; it is cruelty not to do it. Break his will now, and his soul will live, and he will probably bless you to all eternity.
>
> John Wesley, *Works*

Whilst the social climate accepted a high rate of child mortality and suffering, there were moves to improve the lot and lives of the nation's children. Legislation was introduced to restrict and control child labour.

The Factory Act of 1833 forbade the employment of children under the age of nine in specific textile mills and enacted a nine-hour-day for children between the ages of nine and thirteen. The Mines Act of 1842 stated that 'It shall not be lawful for any Owner of any Mine or Colliery to employ any Male Person under the Age of Ten Years'.

Although enforcement of these Acts remained problematical until an adequate inspectorate was established, the nation's children now had legal status. The way was open for further changes and reforms.

At Coxlodge Colliery

Edward Simpson, Aged 13.
Was bad before he went down the pit with (intermitting) fever; and is worse since he went down the pit. Has been down two years and a half. The barrow-men bray (beat) him sometimes.

William Short, Aged about 12.
Is often sickish, sometimes throws up his victuals once a week from foul air, which smells damp-like. Took the measles down the pit. Has been down three years.

John Tulip, Aged 14.
Has been down four years, different pits. They have not agreed with him. His head has worked after, and he throws up when he is out of the pits, twice or thrice a year.'

Children's Employment Commission, Report on the Mines, 1842

'The bits o' lads are badly us'd–
The heedsmen often run them blind–
They're kick'd and cuff'd, and beat and bruis'd,
And sometimes drop for want o' wind.'

Thomas Wilson, *The Pitman's Pay*, 1843

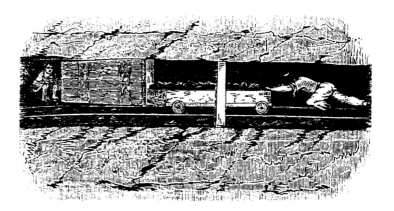

The trapper sits in isolation and often in the dark, opening and shutting the door. The putter is pushing the coal tub from the coal face to a gathering or loading point. Some trappers were as young as five and worked, on average, a twelve-hour day, from four in the morning to four in the afternoon.

The Willington Colliery explosion, 19 April 1841, was caused by a young trapper leaving his door open to play with friends. The ventilation system was disturbed and the build up of gases led to the ensuing explosion. Thirty-two of the thirty-five men and boys below were killed.

Hardship …

The 19th century industrial boom on Tyneside created unprecedented employment opportunities. In some areas, such as Elswick, terraced housing was built to accommodate the rapidly rising workforce. For those who were unskilled, 'the labouring poor', the only homes available were in the already overcrowded slums, as in the Quayside and Sandgate. Here lived many of Newcastle's children. The poverty and deprivation were amongst the worst in the country. In 1895 Dr Thomas Barnardo wrote his 30th Annual Report on the Barnardo homes: 'It is to be noted that I now receive through these provincial agencies some of the very worst and saddest cases of child suffering and destitution which we admit into the Home. Liverpool and Newcastle, for instance, have in this bad eminence, in proportion to their population, far overtopped London, although the latter has so many slums at the doors of its palaces.'

(The first North East 'Ever Open Door' had been opened at 10 Saville Row, Newcastle in 1892.)

In 1897 Dr Barnardo reported again: 'Like all sea ports and great industrial centres Newcastle always has a large number of poor little people whose only refuge in hardship and destitution is the casual ward, the workhouse, or the reformatory'.

No. of Schedule	ROAD, STREET, &c., and No. or NAME of HOUSE	HOUSES Inhabited	HOUSES Uninhabited (U.), or Building (B.)	Number of rooms occupied if less than five	NAME and Surname of each Person	RELATION to Head of Family	CONDITION as to Marriage	Birthday of Males	Birthday of Females	PROFESSION or OCCUPATION
285	5 Picton Ter Chapel Buildings	1		2	William Atkinson	Head	M	38		Cabman Groom
					Elizabeth "	Wife	M		38	
					Eliza Ann "	Dau	S		16	Laundry Maid
					William "	Son		15		Errand Boy
					Dorothy "	Dau			13	Scholar
					Fredrick "	Son		11		"
					Matthew "	"		8		"
					Henry "	"		4		"
					Mary "	Dau				
			1	2	Margaret Stewart	Head	Wid		53	Laundress

This excerpt from the 1891 Census gives an example of the overcrowding in the slum areas of Newcastle. Chapel Buildings, Picton Terrace, was in the All Saints district to the south of Newbridge Street, to the west of the railway line. The three-storey houses were cleared during the 1930s. The Atkinson family, consisting of father, mother and seven children were living in just two rooms. The eldest two children were already at work as laundry maid and errand boy. The father, William, is listed as a cabman groom aged 38. The household is typical of the area.

Sandgate, c.1905.

Playing out

'The street was my second home … When you could crawl and totter you always made for the street whenever the door was open … As soon as you got into that dangerous area, however, some little girl would come to lift you up and totter with you back to safety. They were your street guardians, the little girls.'

Jack Common, *Kiddar's Luck*, Bloodaxe, 1990.

(Jack Common was writing about his pre-1914 boyhood in Heaton but his words seem very apt here.)

Epidemic diseases such as whooping cough, scarlet fever, influenza, diptheria, measles and typhoid raged amongst the children of Newcastle but tuberculosis was the great debilitator and ultimate killer.

'Consumption is most prevalent in All Saint's West, Westgate South, Elswick South, Westgate North, Arthur's Hill, All Saint's East, Byker and All Saint's North … Of all the diseases, this is the one upon which housing conditions exert most influence'.

Harold Kerr, Medical Officer of Health,
Housing in the Quayside Area. Special Enquiry and Report (1913)

'Life in Quality Row', a street in the Lower Ouseburn area, was graphically described in the *Evening Mail*, 3 January 1912. A group of local children were seen as 'a little band of bare-footed desperadoes'. This phrase cannot fit the smiling trio photographed by Mr Halliday. However, the reporter notes that the name 'Quality Row' was assumed to have originated from the 'rich and well-circumstanced people' who first lived in the thirty houses. By 1912 the properties were tenemented and inhabited by more than 120 families. Earlier in the same week 'Mr Coroner Appleby' had drawn the attention of the authorities to the poor living conditions of the families. The humour and resilience of the inhabitants is evident from the concluding part of the article, where the reporter records the following exchange:

> "Weel," says one, "the place isn't eggsackly a pallis, but a lot depends on what we meyk it worsels". "And," remarks a lady, "us is not all heor through wor aan fault, and us divent like taak o' that sort. Thank Hevvings aa hev me pride!" "Gan on," observed a third, "them statements wad meyk ye think us leeved in the slums. This isn't a slum to me. It's hyem."

The light tone of the report belies the lack of positive action by local authorities to provide help for the ragged and barefoot children living on the streets of the city. Initiatives came instead from committed and concerned individuals, either working on their own or supporting national children's charities.

In July 1889 the Prevention of Cruelty to Children Act was passed, a landmark in the history of child care and

*'Quality Row', Byker. Photographed by Fred Halliday for the **Evening Mail**, 1912.*

Playing at shops, Sandgate, c.1898. Lantern slide by Edgar G. Lee.
'On the waste ground at the top, the girls played shop with boody. They pounded up their wares – broken glass and china – with pieces of stone, and arranged them in coloured circles.'
(Arthur Barton, **Two Lamps in our Street,** Hutchinson, 1967. *A memorable book recalling the author's boyhood in the Jarrow of the 1920s.*)

protection. 'The Children's Charter', as the Act was known, allowed the police to arrest anyone ill-treating a child and to obtain a warrant to enter a home where it was suspected that a child was in danger. Guidelines were laid down concerning the employment of children and begging in the streets was out-lawed.

Against this background, the Newcastle Aid Committee of the NSPCC had, by the end of 1890, established a local office and shelter at 57 Lovaine Place. The shelter provided temporary accommodation for children in emergencies. The Inspector, J. McBean, was based at the office but also dealt with cases in most of Northumberland!

In 1897 Dr Barnardo's 'Ever Open Door' in Newcastle had moved to 24 Shieldfield Green. The above extract is from the 29th Annual Barnardo's Report (1894), to indicate the type of case coming to the attention of both organisations.

... and Holidays

A remarkable local venture, the Poor Children's Holiday Association, originated with a letter (right) from John T. Lunn, a Quayside merchant, to John Watson, a clerk and lay preacher, who had been working amongst the poor at the Dunn Street Mission.

This was followed by the first PCHA outing on 11 July, 1891. A group of 120 children were taken to Monkseaton for the day at Mr Lunn's expense. Mr Watson and other Mission workers were in charge. The success of the day attracted public interest and support. More and more holidays followed. The photograph on the right shows children being taken by a Tyne General Ferry Company steamer to South Shields for the day. They were given sandwiches on board and a 'substantial tea' at South Shields.

Gosforth, June 15th, 1891.

Dear Mr Watson,

If there is anyone in your district convalescent or feeble to whom a fortnight's stay at the seaside would be of benefit, I shall be glad to pay for their lodgings, and if necessary, their board as well. Are there any street lads in your Mission to whom a day at the seaside would be a treat? If so, we might organise a trip.

Yours truly,

JOHN T. LUNN

Poor Children's Holiday Association and Rescue Agency,
NEWCASTLE-ON-TYNE (Incorporated).

TRIP TO
SEASIDE
For Poor Children who have no other opportunity of such a day's holiday.
ADMIT BEARER TO
Train, Lunch and Tea,
(For Children 7 to 13 Years).

CHILDREN MUST NOT TAKE FOOD WITH THEM.

This Ticket is only available for the day for which it is issued and is NOT TRANSFERABLE.

Headquarters:
66, Percy Street, Newcastle.

The PCHA movement grew rapidly and in 1912 at least 12,000 poor children had a day at the seaside. Public concern was engaged and other ventures were initiated. The Invalid Children's Holiday Association sent sick children to the countryside, usually for a period of three weeks, sometimes for longer.

To benefit the children excluded from the holidays because of tuberculosis, it was decided to purchase Philipson's Farm, near Stannington, which opened as Stannington Sanatorium in 1907. There will be more about this outstanding development in child health care later on.

Within Newcastle itself, the PCHA opened a Street Vendors' Club in the Prudhoe Street Mission Schools in 1894 to provide food, entertainment and warmth for the hundreds of children trying to earn a living on the streets of the city. Soon it became clear that more was needed. Many of the street children were sleeping out, even in the coldest weather. One boy was found nearly frozen to death, in the portico of the Central Station. In Bottle Bank, Gateshead and Percy Street, Newcastle, night shelters were opened. These in turn led to the Working Boys' Home, in 1895, and a Girls' Home in 1897. Training in the homes was intended to enable the boys and girls to obtain outside employment, when they were old enough to leave.

The letter written by John T. Lunn, in 1891, had started a movement that was to grow throughout the succeeding years and offer hope and comfort to thousands of Newcastle's children. In 1986 the name of the Society was altered to Children North East and its work continues into the 1990s.

A young boy selling newspapers in Grainger Street, c.1898. He would be one of the many to benefit from the Street Vendors' Club. Books, pens, games and paper were provided and impromptu concerts and story telling took place. Hot coffee and a supper would help to keep out the cold.

Poor Children's Breakfasts

There can be no more striking evidence of hunger among Newcastle's poor children in the late-19th century than the newspaper accounts of the provision of free breakfasts. The report below is taken from the *Newcastle Daily Chronicle* of 18 December 1893.

Public support followed as a result of newspaper publicity and throughout the following years local firms, societies and private individuals continued to donate funds to ensure the feeding of the children. The accounts below for the Dunn Street Mission Poor Children's Breakfast Fund, 1893-4, give an indication of this community involvement.

THE BATH LANE BREAKFASTS.

There were 507 poor children present at the free Sunday morning breakfast in Bath Lane Hall, Newcastle, yesterday, and the warm welcome meal provided seemed to be thoroughly enjoyed by all the youngsters present. The Hon. Secretary (Mr. T.M. Grierson, Bath Lane Terrace) desires to acknowledge the gift of 5s from the employees of Messrs. James Smith and Co. On Christmas Day morning a special breakfast will be generously provided at the expense of Councillor Ralph Carr.

In 1892 an engineers' strike caused much distress and hunger prevailed amongst the families of the strikers. Thomas Worthington of Scotswood Road offered to fund breakfasts for 150 children if the Mission Hall in Dunn Street could be used.

Dunn Street Mission Poor Children's Breakfast Fund.

Dr. RECEIPTS AND PAYMENTS ACCOUNT, *Dec. 16th, 1893, to April 5th, 1894* **Cr.**

	£ s. d.		£ s. d.	£ s. d.
To Donations as per lists published in Local Newspapers	102 6 8½	By Special Christmas Morning's Breakfast—		
		R. Embleton & Son, Bread	5 4 0	
,, Cambridge Cyclists' Club, for Sunday Morning Breakfasts to 1,280 Children	5 9 1	J. H. Worthington, Groceries	0 16 5	
		Milk	0 1 6	
		Truttman & Francis, Toys	0 18 0—	6 19 11
		,, Sunday Morning Breakfasts—		
		J. Hume & Son, Cakes	3 4 0	
		T. Pumphrey, Coffee and Sugar	0 18 8	
		T. Worthington, Groceries	0 18 5	
		J. H. Worthington, Milk	0 6 0	
		Cartage	0 2 0—	5 9 1
		,, Daily Breakfasts—		
		J. Hume & Son, Bread	22 6 8	
		Hinton & Co., Cakes	0 15 7	
		R. Embleton & Son, Cakes	1 6 2	
		T. Worthington, Bread and Groceries	58 0 11—	82 9 4
		,, Tea for 250 Parents—		
		J. Hume & Son, Bread	0 19 0	
		T. Worthington, Groceries	0 10 6	
		E. Brough, Hams	1 1 2—	2 10 8
		,, Special Poverty Cases		1 15 0
		,, Cleaning		2 3 0
		,, Coals		0 9 0
		,, Cartage		0 3 6
		,, Sundries		1 3 5½
		,, Printing		1 7 6
		,, Balance in hand		3 5 4
	£107 15 9½			£107 15 9½

Audited and found correct. THOMAS BOWDEN & NEPHEW, Chartered Accountants,
Newcastle-on-Tyne, April, 1894. *Honorary Auditors.*

Emerson's poor children, c.1890.

Robert Emerson, a mineral water manufacturer, inaugurated his new works in George Street, Newcastle, by giving a breakfast for 500 poor children in the basement of the factory. The expressions on the children's faces and the obviously empty bowls say it all.

The *Wellesley* Training Ship

Victorian concern that pauper children would become the future thieves and rogues of British society led to the Industrial Schools' Act of 1866. Schools and training ships were established, part funded by the government and part by private donations, to provide training and education for homeless and destitute boys (unconvicted of crime). The following were the classes of boys to whom the provisions of the Act (Section 14) applied: Any boy …

That is found begging or receiving alms (whether actually or under the pretext of selling or offering for sale anything), or being in any street or public place for the purpose of so begging or receiving alms.

That is found wandering, and not having any home or settled place of abode, or proper guardianship, or visible means of subsistence.

That is found destitute, either being an orphan, or having a surviving parent, who is undergoing penal servitude or imprisonment.

That frequents the company of reputed thieves.

It is important to be aware of the national background to the training ships. Britain, 'the workshop of the world', was at the height of her commercial and imperial power in the mid-19th century. There had been a considerable growth in her shipping but the number of her seamen had not increased in proportion. Here was a chance to train future seamen from an early age.

*The Mess Deck, the **Wellesley**, c. 1895. Mugs, buckets, boys and benches are all straight and ship-shape.*

The following is an extract from a paper given by James Hall (the founder of the *Wellesley*) to the Social Science Congress meeting in Newcastle in 1870:

The general routine of the ship's duties are:- At half-past four in the summer, or five in the winter, the hands turn up, wash, breakfast, and clean decks. At 7.30 in summer, or 8 in winter, divisions for inspection. At 8.30, prayers, then instruction till 11.30. Play before dinner. Dinner at 12. Instruction again from 1 till 3.30, then drill or singing. Supper at 4.30, play, then prayer, and to bed at 7.30; except the night school and reading room party, who remain till 8.30.

A rigorous routine!

The *Wellesley* had been an old line-of-battle ship, the *Boscawen*. In the 1870s, in her role as a training ship, she was moored off North Shields.

The vessel was destroyed by fire in March 1914 and the boys were temporarily rehoused at Tynemouth Palace (more familiarly known as Tynemouth Plaza). Many of the 'old boys' served in the Merchant and Royal Navies and the Army in the First World War. In 1920 the boys were moved to a permanent home in the Royal Naval Barracks on the sea front at Blyth.

The poster (right) for Sports Day, August 1918 indicates the close link between the local community and the boys, confirmed by the many contemporary newspaper reports featuring the ship. It offered hope and a future to boys for whom the prospects were hunger and possible crime on the streets of Newcastle.

'Wellesley' Training Ship Boys'
SPORTS

Commencing
2·15 pm..
SATURDAY,
AUG. 17,
1918.

On the old
ARGYLE
Football
Ground,
HEBBURN-ON-TYNE.

£15 to be distributed in Prizes.

The generous donation of Messrs. Hawthorn & Leslie's Workmen's Allocation Committee.

BOYS will be entertained to TEA through the kind hospitality of the St. Andrew's Institute. ..

24 EVENTS.

A SILVER CUP presented to the 'WELLESLEY' BOYS by the COLONEL & OFFICERS of the 3rd Batt. Durham Light Infantry for BEST SPORTS' MESS of the year will be competed for.

SAILORS' HORNPIPE DANCE by the Boys.

PHYSICAL TRAINING (Swedish Drill) COMPETITION.
(Port Watch v. Starboard Watch   Final).
Mr. Littlejohn's Troupe of Gymnastics will give a Display.

The 'Wellesley' Training Ship Band will be in attendance.

PRIZES PRESENTED by J. T. BATEY, Esq.

These Sports are being held for the Boys as an appreciation to the 'Wellesley' Training Ship for the many services rendered in Hebburn on various Flag Days, &c. The whole of the proceeds will be given to the Boys.

ADMISSION 3d.   **Programme 1d.**

Moore, Printer, N. Shields.

The Dicky Bird Society

One of the most endearing developments for Newcastle's children in the late-19th century was the Dicky Bird Society.

On 7 October 1876 the *Newcastle Weekly Chronicle* opened a column for children, 'The Children's Corner'.

The editor was Uncle Toby, who took his name from 'My Uncle Toby', a character in *Tristram Shandy* by Lawrence Sterne. Hence the 18th century costume for Uncle Toby. Note the small bird sitting on the back of the chair. This was 'Father Chirpie'.

Uncle Toby invited children to join a society, pledging to feed the birds in the winter months and not to rob their nests of eggs in the spring. It was aptly named the Dicky Bird Society. 'The Children's Corner' was a success from the start and children began to write in to Uncle Toby:

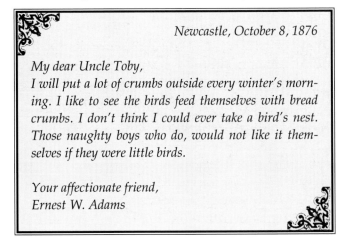

Newcastle, October 8, 1876

My dear Uncle Toby,
I will put a lot of crumbs outside every winter's morning. I like to see the birds feed themselves with bread crumbs. I don't think I could ever take a bird's nest. Those naughty boys who do, would not like it themselves if they were little birds.

Your affectionate friend,
Ernest W. Adams

Within ten years there were 100,000 members of the Society and many were far away from the North East. The first overseas branch was in Norway in 1877. There followed others in Australia, Nova Scotia, New Zealand, Tasmania and South Africa. Children living in many lands were eager to join the

On 26th July 1886 an assembly and event was held in Newcastle to celebrate the DBS membership reaching 100,000. Children gathered in the Town Hall (then at the foot of the Bigg Market) and marched down to the Tyne Theatre. The illustration above shows the procession reaching the bottom of Grainger Street. The plethora of umbrellas indicates the foul weather. However the entertainment at the theatre was such a success that it was repeated three days later.

DBS: 'Indeed it may be said that there is scarcely a district in any quarter of the globe in which English people have settled that does not contain members of the Dicky Bird Society.' ('The History of the Dicky Bird Society', *Newcastle Weekly Chronicle*.)

Uncle Toby asked his young correspondents to mark their letters to him with a drawing of Father Chirpie. The DBS was so well-known that sometimes the drawing and the name Uncle Toby were sufficient address to enable the letters to reach the *Weekly Chronicle*.

Postscript: for many years Uncle Toby was W.E. Adams.

TOYS FOR POOR CHILDREN

UNCLE TOBY'S EXHIBITION OF TOYS, collected for distribution amongst poor children in Hospitals, Asylums, and Workhouses, will be OPENED in the COLLEGE OF SCIENCE, Barras Bridge, Newcastle, on FRIDAY, December 22, at 11 o'clock a.m., and will remain open till 9 p.m.

The MAYOR OF NEWCASTLE (Alderman Quin) has kindly consented to perform the opening ceremony.

The Exhibition will be RE-OPENED on SATURDAY, December 23, at 11 o'clock, and will remain open until 9 o'clock the same evening.

ADMISSION FREE.

Uncle Toby's ventures also benefited poor children. The collection of toys exhibited at the College of Science was for distribution to 'children in Hospital, Asylums and Workhouses'.

Newcastle Daily Chronicle, *18 December, 1893.*

Happier Times … the Seaside

As early as 1807 the Rev. John Hodgson, in his *Picture of Newcastle upon Tyne*, referred to Cullercoats and Tynemouth as places to visit in the bathing season, warm and cold baths having 'lately' been constructed at Tynemouth in Prior's Haven. Whitley he described as a 'pleasant and genteel village'.

The mid-19th century railway boom made the seaside even more accessible for holiday makers. The Blyth & Tyne Railway Co. developed the line to Tynemouth in 1853, the passenger line from Whitley Bay to Tynemouth in 1860 and the Monkseaton link in 1872. The company merged with the North Eastern Railway in 1874, by which time the now familiar circular route between Newcastle and the coast was complete. Commuting from the coast to work in the city was possible and families living in the built-up areas could come to the seaside for the day or longer. The electrification of the line in 1904 made the journey seem even easier. For many children a visit to the seaside became something that dreams are made of.

Plans were underway for the Spanish City Promenade. It was to be constructed, 1908-10, for the Whitley Pleasure Gardens Ltd. and would provide a typical seaside attraction.

Whitley Bay, 1907. The smiles on the faces of the little girls show the timeless delight of sand and sea for children. The large bathing huts to ensure privacy and preserve modesty seem strange now, but were usual at that time.

Tynemouth, the Long Sands, c.1890.

Rowing boats and cautious paddling. Above the Long Sands can be seen the Aquarium and Winter Palace, built 1877-8. According to Tomlinson in his **Comprehensive Guide to the County of Northumberland** *(1888), it contained a collection of: 'birds, beasts, fishes and particularly sea-anemones'. A forerunner of 20th century marinas?*

Sad to say the Plaza, as it was known later, was destroyed by fire in February 1996.

Joe Hind, writing about his Shieldfield childhood in the 1920s and 30s, recalls a summer spent at Whitley Bay:

> During my stay at the seaside, my cousin and I shared an attic bedroom with a tiny window that looked over the North Sea. I would sit and watch for hours the huge revolving lamp of St Mary's Lighthouse, casting its light over the vast expanse of the sea.
>
> *A Shieldfield Childhood*, Newcastle City Libraries, 1994

For Jack Common and his friends, coming from the terraces of Heaton, an expedition to the seaside was an adventure. They made their way mainly on foot, having only the fare for the tram from Wallsend to North Shields.

> Walk? We often ran. Why, here from North Shields on, the air was full of the great sea-glow, a salt radiance brightened all the long Tynemouth streets. And at the end of them, the land fell off at the cliff-edge into a great shining nothingness immense all ways over the lazy crimping of seas on their level floor.
>
> *Kiddar's Luck*, Bloodaxe, 1990

... and the Parks

In 1845, the admirable Dr Reid stressed the importance of open spaces within and close to the town for fresh air and relaxation. It took time for his ideas to take root but by 1873 pressure by Alderman Sir Charles F. Hamond led to the conversion of 68 acres in Castle Leazes into Leazes Park. This was Newcastle's first park.

Between 1878-80 more areas of the Town Moor were laid out by the Corporation as parks, including Exhibition Park and Brandling Park. As this was common land it could only be converted by agreement between the Freemen and the Corporation.

In 1880, Lord Armstrong gave Armstrong Park to the town (Newcastle did not become a city until 1882). Shortly afterwards he generously added Jesmond Dene and the Banqueting Hall. With the opening of Heaton and Elswick Parks Newcastle had more open spaces reserved for public recreation than any other town in the country.

The parks provided bowling greens, band-

stands, lawn tennis courts, quoits grounds, croquet lawns, flower gardens and lakes. It may seem that these benefited only the middle classes but the parks provided open spaces for children from the crowded areas of Newcastle too.

The Lake, Leazes Park, c.1900.

Doris Chambers, writing about her 'Ordinary Childhood' in the West End of Newcastle in the 1920s, gives an enthusiastic impression of the parks:

> The parks gave us great pleasure, they were beautifully maintained with the flower-beds a riot of colour and every corner utterly immaculate. The tennis-courts were always very busy with would-be stars, the bowling greens like velvet, and the model house was an everlasting source of great joy to me with its beautiful statues.
>
> Doris Chambers, *An Ordinary Childhood*, 1973
> (unpublished)

Mrs Chambers' reference to the 'model house' and the 'statues' may refer to the collection of sculptures by J.G. Lough, housed in Elswick Hall. Elswick Park was her nearest park and the Hall was open to the public.

Benwell Park was given to Newcastle by Dr Thomas Hodgkin, the banker, in 1899, and it became better known as Hodgkin Park. It was opened with considerable ceremony in 1899. There was a procession, which started from Adelaide Terrace, Benwell. Free food and drink were provided in the park, while a band from Byker played.

Postcards of Hodgkin Park in the 1900s show waterfalls, woodland walks, lakes, bowling greens and fountains. For families living in Newcastle's West End it must have been a delight.

Hodgkin Park, c.1905.

The boy and girl in the photograph are dressed in their best clothes and look suitably serious. The fashion for muffs, neat boots and wide-brimmed hats is shown in the girl's stylish outfit. The boy, with his Norfolk jacket, breeches and walking stick looks the epitome of a young country gentleman.

Home Sweet Home

These photographs give glimpses of home life for children living in more comfortable surroundings than those described in the earlier part of the book. The photograph on the left shows Gloucester Terrace, Elswick, in 1912.

The McWilliam children stand by the front door to their home in Gloucester Terrace. The grown-ups in the family were well-known butchers in Scotswood Road and cattle market wholesalers. The house is substantial, with basement rooms and attics. Net curtains and blinds hang at the windows. The children's white collars show a determination to maintain a high standard of cleanliness and to keep up appearances. This would be achieved through much hard work by the mother and in this case, possibly, the maid. Washdays were a mammoth operation involving poss tubs, poss sticks and mangles. Clothes lines were suspended across back-lanes and a sharp lookout had to be kept for passing tradesmen, especially the coal cart. It was not unknown for a horse and cart to be driven through the lines of washing, which would mean a complete rewash.

Undoubtedly the stone steps to the McWilliams' house would be cleaned every day with a rubbing-stone. Tom Callaghan, describing his Benwell childhood in *Those were the Days* (Newcastle City Libraries, 1992) writes about a 'Mrs Rubbing-Stone', though her steps sound less prosperous than the ones shown here:

> She accosted me as I was on my way down to the terrace to shop around for bacon-pieces and she asked me to bring her a rubbing-stone from Storey's, the hardware shop: 'Now this is the colour I want,' said she in

determined manner and holding up the stone she was using at that moment. 'You know what colour this is don't you?' she asked decisively. 'It's the colour of ginger-bread, Missus,' I responded. She smiled, looking pleased at my appropriate description.

The photograph on the right, belonging to Desmond Walton, is of his maternal grandparents, Robert and Statira Hisco. They are with their children, Ralph and Lizzie, outside their home in Westerhope. Robert Hisco originally came from Kendal, moving to Newcastle to set up as a joiner and builder in Northumberland Road. He became a member of the Northern Allotment Association, which was formed to buy land in rural areas away from city smoke and to construct smallholdings. The Association bought the Red Cow Farm, East Denton, originally part of the Montagu Estate and close to Stamfordham Road. Robert built one of the first houses on the estate. The name of the area was changed to Westerhope in 1896, meaning that those involved in the venture were coming west with hope for a new life.

The Hisco family at home in Westerhope in 1896. Ralph Hisco sits astride a rocking horse made by his father. Behind him the French windows and the potted plants indicate the level of comfort in this new home in the west.

25

A group of children playing at school, a serious teacher and an embarrassed dunce. The 'class' is enjoying the joke, back-lane, Ancrum Street, Spital Tongues, c.1920s.

'We had to make all our own entertainment and we certainly managed to do so. The back-lanes, of course, were our playground. The older boys had grooves worn into the earth between the cobbles in which they played 'three-holes' with their favourite marbles. ... Then there was 'up for Monday' – a game in which we had to throw a ball onto the lowest roof at the back of someone's shop and catch it as it came down … We would often play at 'tredgy' or 'rounders'. Some of the girls could throw the ball almost as far as the boys – no mean feat. 'Kick the block' must have had the residents of the houses at screaming point. This game was an old tin kicked by each player in turn and the din must have been somewhat stupefying.'

(Doris Chambers)

The list of games continues with some unusual names such as 'hitchy-dobbers', a form of hop-scotch in which the pavement was marked into squares onto which a pebble was thrown and along which the players hopped. It certainly seems that the lack of modern sophisticated toys did not stop Newcastle's children of past years from enjoying themselves in a total and wholehearted manner.

Sunday Best

Churches of all denominations provided the focal point for the social lives of most Newcastle families. Children were involved and catered for from a very early age. Local autobiographies and oral reminiscences recall the impact on childhood of church attendance, the Sunday schools and the tradition of the Sabbath.

Sunday was the day to wear your 'Sunday best', primarily for going to church. Those who did not have a 'Sunday best' felt that they could not go to church, comments Arthur Barton sadly. He did not feel that Sunday clothes were constricting, but it was, instead, a pleasure to wear something smart. Sometimes the children would be given a buttonhole of flowers to wear, a really special touch.

The Sunday meal was the highlight of the week on the culinary side. Jack Common writes about having 'half a cow, greens, potatoes and Yorkshire, and lashings of rice pudding to follow'.

After this feast the children left for Sunday school while their parents went to bed: 'Perhaps one reason why the working-class in general is so good-humoured and patient, charitable and unenvious is that they must have been, a great lot of them, conceived on the day of grace.' (Jack Common).

Sunday schools had been established in the 18th century and offered education for the working children of poor families, Sunday being their only free day. Prizes, usually books of a religious nature, were given for good attendance. In the summer, boat trips along the Tyne, train excursions to the seaside

St George's Catholic Church, Bell's Close, Scotswood, 1908. Here lived many Irish Catholic families who had come to work in the riverside factories. The spotless white pinafores protect the girls' clothes and the boys wear the cloth caps seen in so many pictures. The end of a terrace of working class housing can be seen. Mothers would struggle hard to make sure their children looked smart.

Sunday schools' procession, Good Friday, Benwell, 1910.
The procession is passing through Strathmore Crescent. Note the variety of hats.

and picnics in the country were organised. On these outings each child was provided with a bag of food and for many they were days to treasure and remember. Christmas was a time for treats, teas and simple gifts. For poor children the church festivities were often their only glimpses of Christmas.

As better-off children returned home from Sunday schools the Sunday tea was the next feature of the day. Here, home cooking came really to the fore. There would be several sorts of scones, sausage rolls, jam and fruit tarts and a variety of cakes.

In the summer the day might end with a walk in a nearby park. In the winter there was the attraction of a magic lantern show. Joe Hind recalls the lantern shows at the Henry Street Mission in Shieldfield. The underlying message of these entertainments was to warn children of the evils of drink.

Whit Monday procession of witness, Walker Presbyterian Mission Hall, c.1905.

Decorated floats, known as 'Rollies', carrying the children of the primary Sunday school classes, formed part of the procession of witness by the Protestant churches of Walker on Whit Monday morning. Each church would decorate one or two wooden frames and these would be fitted to the flat carts belonging to street traders. There was rivalry to produce the most original and best decorated 'Rolly'.

The children met first at their own chapels and were given a bag containing fruit, cakes and a few sweets. The older children walked behind the 'Rollies'. The stronger of the boys would carry the church banners, the girls held the guy ropes. Teachers, parents and helpers joined the procession and short services of witness were held along the route. The procession ended at the far end of Church Street where there were fields. It was time for a picnic and games.

W. Muir, *Memories of Walker* (compiled for Wharrier Street Junior School Golden Jubilee, 1934-84.)

School Days, 1870-1914

The Board schools (established by the Forster Education Act, 1870) signified a major commitment to education by the state. Previously, voluntary societies had been relied on to provide and run the schools. On a local basis the Board schools made a considerable impact. Newcastle was divided into four districts and large, imposing schools were built in each. They had two or three entrances, one for the boys, one for the girls and one for the 'mixed infants'. Board schools were allowed to impose a weekly school fee until the Free Education Act of 1891. Compulsory attendance was enacted in 1876 and the 'School Board Man' would visit the homes of truant children.

Todd's Nook School, in Arthur's Hill in the west end of Newcastle, was opened by the Lord Mayor in 1891. At the time it was the largest school built by the Newcastle Board. The curriculum concentrated on 'Biblical instruction and the principles of morality, reading, writing and arithmetic, English grammar and composition and the elements of geography and history'. ('Scheme of Education' drawn up by the members of the first School Board, 1871.) To vary the routine, 'object' lessons were a feature of the school days. A piece of fruit or a plant would be displayed and the class told of its origins and composition. Music was taught by learning the Tonic Solfa, which must have stifled much joy. For girls, cookery and housekeeping classes prepared them for work 'in service' and motherhood. The serious faces of the children in the photograph of Todds Nook School, pictured in 1897 on the front cover of this book, suggest a disciplined and controlled atmosphere in the school room.

Chillingham Road School, Heaton, c.1910.
The school celebrated its centenary in 1993.

I have to report the death of one of my little Scholars. He was at School last Wednesday apparently quite well and full of life. He complained of being tired on Thursday Morning so his mother sent me word to that effect, but that he would be at School in the afternoon. He did not come but played at home & went some messages.

On Friday morning he died in Convulsions and was buried yesterday. He was nearly five years old & resided at 122 Monday Street. His Teacher has called upon Mrs. Fraser this morning.

A poignant extract from the Log Book, Todd's Nook School, 11 May 1897.

The girls in the photograph below are sitting sedately with their hands on their laps. Exercise books are on the desks but the class used to practice their writing on slates. Behind the teacher are screens which could be moved to alter the size of the rooms. The 1902 Education Act had abolished the School Boards and instead created 140 Local Education Authorities, LEAs, and the Board schools became Council schools.

These photographs are of 'provided schools', which by 1914 numbered thirty-six in Newcastle. Alongside them were a number of schools run by voluntary societies and private organisations. Many of these were Anglican or Roman Catholic but there was also the Bath Lane School founded by Dr Rutherford, the Ford Pottery School at Ouseburn, a Jewish School and the Elswick Works School for the children of employees at the Armstrong Works, Elswick.

Childhood ended comparatively early. After the age of 13 many of Newcastle's children became contributors to the family budget. To be an errand boy was often the first step into adult work. There was a background of earlier experiences, such as jobs to be done on Saturdays to help parents. The week's supply of coal had to be collected in a barrow, and a plentiful number of freshly baked loaves in a white pillow case.

Most boys hoped to have had a paper round and the deliveries might take them into the growing wealthy suburbs such as Jesmond and Gosforth.

Elswick Road School, 1914.

The Royal Victoria School for the Blind, Benwell, c.1905, a rather moving photograph masking some of the considerable advances being made in the education of blind children.

In 1838 the 'Northern Asylum for the Blind and Deaf and Dumb' was opened in the Spital area of Newcastle, near Westgate Street. By 1841, the location of the asylum had moved to Northumberland Street and the name was changed to 'The Royal Victoria Asylum for the Blind'. 1893 was a significant year. The asylum committee bought the house and grounds of Benwell Dene from Dr Thomas Hodgkin (of Hodgkin Park fame) and the Elementary Education (Blind and Deaf Children) Act was passed. More thought was to be given

to providing a wider curriculum for blind children, away from the purely basket making and trades-based training.

This was emphasised by the dropping of the name 'asylum'. In 1895 the institution became The Royal Victoria School for the Blind. By the time of this photograph the blind children were being taught nearly all the subjects included in the curriculum of an ordinary elementary school.

The Great War

Family life was deeply affected by the outbreak of war in August 1914. Fathers, brothers and uncles left for the front, many never to return. The streets of Newcastle were lined by children waving their Union Jacks as the soldiers marched past.

It was left to the women to keep the home fires burning and to ensure that the shipyards and armaments works met the demands of the great crisis. The lack of nursery care meant that those with young children had often to leave them unattended, to survive on the streets. Schooldays for the older children were sometimes disrupted:

'On account of War with Germany only morning school is being held. The infants from Wingrove School are occupying this department in the afternoons, as their school is in the hands of the Military Authorities.'

(25.8.14, Log Book, Todd's Nook School.)

The Log Book continues to record the collection by the children of money for the Belgium Relief Fund, the First Northern Hospital, the Lady Mayoress' Appeal and the Queen's Appeal. Socks, handkerchiefs and postcards were purchased to send to the soldiers at the front and in the military hospitals.

By 1917 the attacks by German U-boats on allied shipping led to grave food shortages, with queues in Newcastle to buy groceries.

Despite these years of hardship, after the Armistice was signed in 1918 mammoth efforts were made to celebrate in style. Victory street parties were held throughout the city.

Victory tea party, Clifford Street, Byker, 1919.

The flags are flying, including Japan's, our ally in the war. There are few young men present and the celebration is tinged with sadness. The young girls favour dressing up in the uniform of a Red Cross nurse.

Life on the Dole

The relief and rejoicing at the end of the First World War were tempered by alarm and concern. In 1920 there was an interlude of industrial activity on Tyneside whilst the shipyards worked to replace the wartime losses, but by 1921 virtually all the shipping orders had dried up. The years of high unemployment and hunger began. The seven-month long Miners' Strike, 1926, and the General Strike, 3-12 May of the same year, were followed by the hunger marches of the 1930s.

Andrew Barton, in Jarrow during the Miners' Strike, remembered boyhood forays to search for anything that would fuel the family fire. Policemen stood guard over wooden railway bridges. His father, a crane driver, began a long period of unemployment and family life was strained. In Newcastle, during the Thirties, 30,000 men were on the dole. The infant mortality rate was 91 per 1,000, whereas nationally it was 53. The rate of tuberculosis was 40 per cent above the national average.

11 & 13, Fell Street, Byker, c.1935.
Families are standing outside their homes in Byker. For them, living conditions had changed little since the end of the 19th century.

… New Homes with Gardens

In 1930 nearly a quarter of Newcastle's inhabitants were officially classed as living in overcrowded conditions, compared with less than a tenth nationally. Between 1890-1920 Newcastle Council built only 454 homes, but during the 1920s and particularly after the 1930 Housing Act, a commitment to new housing emerged. Slums were cleared and in the following decade 8,130 council houses were built.

Maureen McDade (interviewed March 1996) recalled her family living, from 1933, in a new council house in Cranbrook Road, West Benwell. Her father was a welder at Vickers Armstrong works. There were eventually six children in the family. The three-bedroomed house 'always had a bathroom', 'there was always hot water' and a separate toilet. At the front and back were gardens.

By 1939 the city had provided homes for more than 22,000 people.

Clearly this is an idealised drawing, but it shows the move away from the 'two-up, two-down' homes. Here are gardens and space for Newcastle's families.

*(From **Newcastle upon Tyne Housing: a Brief Review of Post-War Activities**, 1926.)*

Boy Miners

At the front, left hand side of the photograph below is Cud Smith, aged fifteen, the youngest of the boy miners leaving the Charlotte Pit. Cud was earning 2s 4d a day. The other boys were taking home 3s 4½d a day.

The boys look cheerful enough but the reality was a tough, harsh life. On 30 March 1925, at the nearby View Pit, Scotswood, occurred the Montagu Pit Disaster in which 38 men and boys died when water burst into the mine from older workings.

After the protracted strike of 1926, the starving miners returned to work for longer hours, at lower wages, than when the dispute first started.

Bill Johnstone, in *Coal Dust in my Blood*, (Oolichan Books, 1993), recalled being taken at the age of thirteen by his father to Ashington Mine Office. The year was 1921. By law he could not go underground until he was fourteen, but he could work on the picking table for 10d a day: 'Our job was to pick out the pieces of shale from the coal as it passed on a conveyor. The boredom of watching a slow-moving conveyor passing one's eyes for eight hours a day, six days a week, was enough to drive one crazy.'

After a year Bill went down the pit and became a trapper: 'My job was to open and close a trap-door to allow a set of tubs pulled by a pony to pass through.' This recalls the description of a boy trapper on page 7. Apparently little had changed.

Charlotte Pit, Benwell, June 1929.

36

The Young Pioneers

One of the most disturbing and distressing chapters in British history concerns the migration of children from British orphanages and institutions, c.1860 to 1967, to the dominions and colonies.

Prior to the Second World War migration occurred against a background of poverty and deprivation in British cities, an anticipated life of crime for the destitute children and, in contrast, the supposed opportunities in the underpopulated colonies. The reality for the migrant children was often a nightmare.

From about 1880 to the 1930s Canada was the principal destination. The boys went to work on farms and the girls were employed in domestic work. On arrival in Canada they were often sent to remote areas and although some were well-treated, many were exploited, underfed and abused. Newcastle's children were among the migrants.

The Annual Reports of the Boys' Refuge, 204 Westmorland Road, record that five boys went to Canada in 1924, one to Australia in 1926 and one to Canada in 1927.

On 13 April 1910, a group of boys from Philipson Farm Colony, Stannington, left for Canada. They travelled under the care of Salvation Army officials, who were to place them on farms 'where it is confidently believed they will become good and capable citizens.' (Charles Wain (ed.), *A Romance of Regeneration*, Andrew Reid & Co., 1913).

In 1911, fifteen more boys left for Canada from the Farm Colony.

In the *Evening World* for 28 September 1929 a headline reads 'Training the Young Pioneers'. Boys between the ages of fourteen and nineteen were being taught farming work at Walker Farm School before migration to Australia and Canada.

After the Second World War there was a further increase in migration. British orphanages were full and social services were over-

Philipson Farm Colony boys starting on their journey to Canada, 1911. The location of the station is not known. The boys look well-dressed and equipped for the journey.

Walker Farm School, 1929. Boys are in the school's reading room, endeavouring to be serious.

Australia to a virtual slave camp, Boys' Town Bindoon.'

In his book, *Geordie: Orphan of the Empire*, (P & B Press, Perth 1990), Lionel describes his experiences in both the orphanage and the Boys' Town. The harshness of the regimes, particularly in Bindoon, profoundly affected him as an adult and is remembered with great bitterness. He only learnt the identity of his parents when, at the age of fifteen, he accidentally saw a Registration Book at Bindoon. At two years old, he and his sister (aged six months) were placed in the care of Roman Catholic nuns by their mother, who had contracted tuberculosis. Their father had deserted the family. Both the nuns at the orphanage in Carlisle and the Christian Brothers at Bindoon had concealed Lionel's true identity from him.

Many children were sent abroad from adoption homes without parental consent, and once abroad were discouraged from trying to contact their families. In many cases they were placed in large institutions, sometimes in remote areas and some suffered cruelty and abuse. The last group of child migrants left Britain in 1967 for Australia.

The Child Migrant Trust has been established in recent years to try to trace relatives of migrants, so reunions can be arranged before it is too late.

stretched. The Australian government was concerned about the country's low population. There followed a large scale exodus of children (including some from Newcastle) from homes in Britain to Australia. A few were as young as four years of age.

Lionel Welsh, living in West Australia, in a letter to the author, April 1996, wrote: 'I was born in Sutton Dwellings, Newcastle, 1936, hijacked to Carlisle in 1938 and shipped to

Brighter Futures

The dark days of the 1920s were cheered by two major landmarks in the history of Newcastle: the building of the Tyne Bridge, 1925-8, and the North East Coast Exhibition, 1929. The children of the city were involved in the celebrations and activities.

The *Evening Chronicle*, 10 October 1928, reported the visit of King George V and Queen Mary to open the bridge and to open the new Heaton Secondary Schools. One headline read: 'Massed Youth welcomes their Majesties at Heaton. 24,000 Schoolchildren Sing Heartfelt Greeting to Royalty'. The royal visitors were treated to 'a display of physical drill and country dancing given by secondary and elementary schoolgirls'. The King responded: 'It is a great pleasure to us to have this opportunity of seeing so vast an assembly of schoolchildren and of visiting the new Heaton Secondary school.'

The headlines for the ceremony at the Tyne Bridge read: 'King's Faith in Men of Tyneside. Royal Hope that Bridge Means New Tide Of Prosperity.'

On 14 May 1929 the Prince of Wales opened the North East Coast Exhibition. The pavilions and stadium were sited on part of the Town Moor. Young people's organisations took part in parades and gymnastic displays as part of the Pageant of Youth. The exhibition attracted over 4 million visitors and showed that, despite the Depression, the resilience of the North East offered hope for the future.

As well as opening the new Tyne Bridge, King George V and Queen Mary visited the new Heaton Secondary Schools on 10 October 1928. The Schools accommodated 500 boys and 500 girls.

*En route to the Tyne Bridge the Royal couple also passed the boys of the **Wellesley** training ship, Axwell Park School and the children of the Northumberland Cottage Homes.*

The King and Queen at Heaton Secondary Schools, Oct. 10th, 1928
N/c. Evening Chronicle Photo

Following in the footsteps of the Dicky Bird Society came the Gloops' Club, inaugurated by the *Evening Chronicle*, 2 January 1929. It was called after a cartoon cat, Gloops, 'the funniest cat in the world'. This time the central figure was Uncle Nick of Radio 5NO. He was supported by Aunt Ann. The motto of the club was 'Thmile' (Gloops always spoke with a lisp).

Members would be able to take part in the *Chronicle*'s 'Sunshine Fund' for poor and sick children. The fund had been inaugurated by a donation from King George V.

Following editions of the newspaper and the Gloopers' own supplements gave details of charabanc outings and concert parties. (An assembly of Gloopers was a Gloopege.) There were drawings and instructions on how to make such useful articles as a comb case or an egg cosy. ('Blue Peter', here we come.)

Medals were awarded for 'heroism, scholastic successes, school records, athletic prowess and acts of kindness and self-sacrifice'. Uncle Nick's dog Jack, an Airedale, and two Shetland ponies joined the team. It was obviously a most successful venture and by August 1929 the membership had reached 110,000.

Dear Boys and Girls,—The Twins have had the chance to pick some fruit in a kind friend's garden, so this morning they decided to make some jam, and put the fruit to stew with some sugar on the stove, and while they went to get the jam pots they told the Glooper to attend carefully to the

JAM MAKING.

GLOOPERS' JOY DAY.

Happy Alnmouth Trip for 120 Children.

One hundred and twenty members of the "Evening Chronicle" Gloops Club left Newcastle yesterday to spend a happy day at Alnmouth.

They were in the care of Cousin Rene and six of the "Evening Chronicle" pages, and travelled by omnibus.

The trip was provided specially for children who had not yet had a holiday nor any prospect of one.

Uncle Nick saw the children off, and they greeted him with rousing cheers.

'The Gloopers Own' headline and the extract, above, are from the **Evening Chronicle**, *19 September 1929. The cartoon, left, from the same date, is one of many featuring Gloops and the Twins.*

Nursery days

The 1920s and 1930s saw major developments in the care of the youngest children. During the First World War Miss Greta Rowell was disturbed by the number of children roaming the streets of West Newcastle while their mothers worked in the armament factories. In 1917 Miss Rowell persuaded Mr Frederick Milburn to buy 33 West Parade, on the corner of Westmorland Road. This became the West End Day Nursery. Mothers paid eightpence a day to leave their children from 5.40 a.m. to 7.00 p.m. Later they paid just sixpence. Soon the nursery was open at night for mothers working on night shift.

Dr Rutter, the Nursery's medical officer, reported his examination of 146 children in 1918: 'The frequency of improper feeding in the homes … the lack of care in keeping the children reasonably clean … the number of children in whom diseases such as rickets are neglected … Children need care and attention as well as love.' (Ursula Ridley, *The Babies' Hospital*, Newcastle upon Tyne.)

It was vital to monitor the health of the very young to prevent the development of serious medical problems and the importance of providing more than just practical care was now recognised. The correct psychological treatment of children became important. Dr James Spence (see page 47) had joined the staff of the Nursery and his influence was being felt. His work in Newcastle was to revolutionise the care of children in hospital both locally and nationally.

In 1925 the Day Nursery became 'The Babies' Hospital and Mothercraft Centre' and remained so until it was amalgamated with the Royal Victoria Infirmary in 1944.

In 1932 Newcastle Education Committee opened Ashfield House, Elswick Road as a nursery school and clinic. This was the first local authority nursery school in the North of England. Elswick was severely affected by the economic depression and its children were suffering. Here was a venture to help the local community. Ashfield House was the former home of the Richardsons, the owners of the Elswick Leather Works.

During the Second World War Elswick and its factories were a prime target for bombing raids and the school was evacuated for a short time. By 1941 the need for nursery care was vital for the mothers, who were again providing the workforce for the munitions factories. Ashfield House opened as a war nursery.

A relaxed photograph of a group of youngsters sitting at the back of Ashfield House with Miss Jaqueline Gibson. Miss Gibson joined the staff in 1947 and remembers it as a 'happy and peaceful place'.

World War II and the Evacuees

The outbreak of World War II in 1939 disrupted the lives of British children. For many it came to mean separation from their families and unsettled schooldays.

As Europe had edged towards hostilities in the late 1930s, the fear prevailed of large-scale bombing raids on civilian targets.

The British government, to prevent a breakdown of civilian morale and panic flight from the cities, decided to evacuate children and mothers with babies to the countryside. The country was divided into evacuation, neutral and reception areas. The reception areas were to billet the evacuees. Evacuation was voluntary. Rehearsals were held throughout the summer of 1939 so that when Hitler invaded Poland on Friday 1 September the evacuation scheme went into effect immediately. Within a week, one and a half million people were evacuated. Another two million had made private arrangements to leave for safe areas.

On 1 September 31,222 Newcastle schoolchildren were evacuated. The following day a further 12,818 mothers and children left the city. A *Newcastle Journal* reporter described the scene at the Central Station:

> The children were like "little Soldiers" with their equipment as they stood about. Some of the older boys had laid their knapsacks on the platform as a pillow and were lying reading papers for all the world like old campaigners.
>
> When the trains steamed in there was no rush – school discipline was strong and the children filed in and everyone obtained a seat. Mothers appeared to have obeyed the injunction not to go to the station.

Evacuees, September 1939, St Joseph's School, Armstrong Road, Benwell.

The group of evacuees is at Egremont, Cumberland for a temporary stop. They were travelling by train from Newcastle to Whitehaven and on to Cleator Moor, by bus. The photograph belongs to Terry Quinn, who is standing on the far right, with his mother and younger sister Sheila. (Terry was 11 years old at the time.) After six weeks the Quinn family returned home. It seemed in the months of the 'Phoney War' that the fear of bombing raids would not be realised.

The national newspapers echoed this positive view. The operation had gone like clockwork and was a great success. However, for many of the children and mothers it was a heartrending time and when there appeared to be no immediate danger of bombing a number of children were brought home.

Back in Condercum Road, Benwell Terry Quinn found his school closed and began his own private contribution to the war effort. He collected sandbags and gas masks from the local council depot and delivered them to nearby houses. Sometimes he was rewarded for his efforts.

With the occupation of France in June 1940 and the perceived threat of invasion, a second evacuation scheme was organised. 4,300 Newcastle children left the city by train.

The Royal Grammar School had prepared carefully for evacuation to a country house in Penrith but experienced the familiar evacuation problems of enuresis (bed-wetting) and home sickness. Ian Atkinson remembered that a few boys tried to run home 'with six pence in their pocket and a school atlas'. A 'posse' of masters would set off in pursuit. These events were the exception. The school was aware of the need for pastoral care. Hostels were provided for the boys in addition to billets. Ian was in Lynwood hostel and enjoyed his years in Penrith. The school joined in the life of the town. The boys participated in productions at Penrith Playhouse, helped on local farms to lift potato crops and cultivated an allotment.

The Central Newcastle High School was evacuated to Keswick on 1 September 1939. 'On arriving at school at about 1 p.m., the first thing which greeted us was a large notice in the entrance hall, "Do not drink water in trains. It is bad."' (Doreen Murray, quoted in *Central Newcastle High School, 1895-1995*.) The school stayed at Barrow House hostel.

Girls helped with haymaking and knitted garments for the forces and bombed-out families. The continuity of school life and contact with Keswick school enabled the pupils to keep up with schoolwork and take the school certificate examinations.

TAKE THEM BACK! TAKE THEM BACK! TAKE THEM BACK!..

DON'T do it, mother—

LEAVE THE CHILDREN WHERE THEY ARE

ISSUED BY THE MINISTRY OF HEALTH

A government poster dating from the 'Phoney War', warning mothers against the temptation (personified by a shadowy Hitler) to take their children home, away from the safety of the countryside.

A Land Fit for Children

In 1944, Barnardo's were accused of providing poor education for the under-fives. A Barnardo book was published to provide guidance for the homes. For example, children should not rise before 6.30 a.m. and in the dining room complete silence was 'not desirable as it savours of the Institution rather than the Home'.

The 1948 Children Act specifically required local authorities to set up departments for the care of destitute children, including the provision of residential homes. This gave Barnardo's the freedom to develop more support for children and their families within the community.

The 1948 Act was based on the recommendations of the enlightened *Report of the Care of Children Committee*, September 1946, chaired by Miss Myra Curtis. The report included a section on 'After Care': 'This should be a matter of deep concern. Great care should be taken to make the children aware of the possible careers open to them … Any deprived child going into employment in a strange place should be enabled to get into touch with the Children's Officer of that area.'

Hopes for the future were high.

Dr Barnardo's Home, Millbrooke, Scrogg Road, Walker, 1947.

The background of roundabouts and swings shows that the care of children in residential homes was improving. In 1941 Barnardo's had established a training school for its staff: 'Students studied child care, child psychology, hygiene, first aid and home nursing, children's hobbies, games and Bible story telling.'
(**Barnardo's Children**, *Barnardo's, 1995.*)

Getting Better and Staying Well ...

The years covered in this book saw some remarkable and essential advances in the health care of Newcastle's children.

At the beginning of the 20th century, the city had one of the worst national averages for infant mortality and deaths from tuberculosis. The causes lay in malnutrition, overcrowded and poor housing and industrial pollution.

Improvements came through the work of private organisations, the local authority and a number of exceptional individuals dedicated to the care of children.

Stannington Sanatorium opened in 1907 as a result of the work of the enterprising Poor Children's Holiday Association. By 1926 the number of beds had risen to 312 and in 1932 a school was opened for the children.

The hospital buildings were remarkable. The ward verandahs were fitted with 'Vita' glass (dictionary definition – 'kind of glass by which the ultra-violet vitalizing rays of sunlight are not excluded as by ordinary glass'). The operating theatre and the X-ray department were of the most modern design.

In the 1960s the treatment of tuberculosis radically changed with the introduction of new drugs and the sanatorium became a convalescent hospital for children recovering from serious and long term illnesses.

The Open Air School, Stannington, 1934
This optimistic photograph shows children enjoying the fresh air and sunshine, whilst continuing their education.

Postscript. An extract from the *Newcastle Journal*, 25 March 1996:

'Last year, 35 cases of the illness' (tuberculosis) 'were recorded in adults in Newcastle and North Tyneside alone – more than double the lowest figure of 16 in 1991. In Newcastle, a definite link has been noted between TB cases and poorer socio-economic groups – particularly among the homeless or those living in hostels.'

Children in Hospital

The Fleming Memorial Hospital was opened in 1888. The funding for it was provided by John Fleming, a Newcastle solicitor. The hospital, built as a memorial to Mr Fleming's wife, was adjacent to the Town Moor, a more open and healthy environment than the old Hospital for Sick Children in Hanover Square.

The wistful face of the patient in the photograph below prompts the question – 'What was it really like for the children in hospital?'

In 1921 James Perry of Gosforth, aged six, collapsed with peritonitis. He had become ill through the night and the doctor, realising the seriousness of the case, took him into the Fleming. James was operated upon immediately. The operation took from 8 a.m. – 11 a.m. Recovery was long and difficult. There were a number of relapses and the nursing staff contacted James' father several times. It was thought that James was dying. His parents and his four year old sister were allowed to stay overnight during these crises. This was exceptional because the visiting hours, 6 p.m. – 8 p.m., were usually adhered to strictly. For James the experience was traumatic and he lost six months of schooling. Whilst he remembers the rigid discipline of life on the ward, he was well-cared for and survived the experience.

For the nurses the discipline was equally severe. Miss Hilda Hope was nursing at the Fleming from 1935. She recalls that the nurses had only one day off a month and they had no choice as to which day it would be. Miss Hope saw advantages in the restricted visiting. On one occasion the staff were having trouble 'stabilising' a little boy with diabetes and discovered that his family were bringing him unsuitable food. In the years prior to antibiotics, when hospitalisation was usually a lengthy process, nursing care was all important in the recovery of sick children.

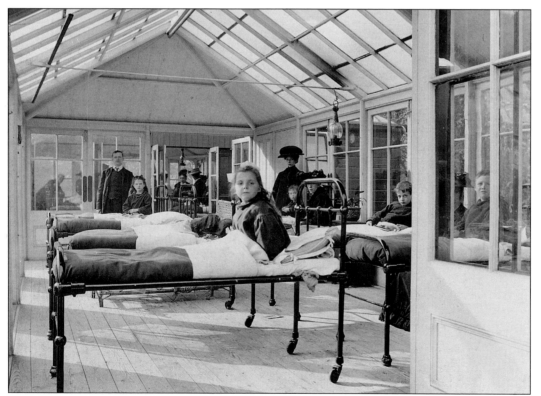

The Fleming Memorial Hospital, pre-1914.

Maureen McDade was admitted to Walkergate hospital with meningitis in 1944. She had become so ill at home that a neighbour had come in with sheets to 'lay her out'. Maureen remembers waking up in hospital with her father sitting next to her 'with a big white handkerchief'. Again, visiting was allowed because of the serious nature of the illness. Maureen was later told 'the lines were out for you'. This referred to the practice of publishing information about patients in the *Evening Chronicle*. A number was given to those admitted to hospital. The relatives would scan the newspaper looking for the number under the relevant hospital. Categories of illness would be listed: 'Dangerously ill, Very ill, Slight improvement, Improving, Much the same'. In the days when telephones were few, this was the best means of communication for the families of sick children.

In the 1930s Norman Pallace, suffering from scarlet fever, was a patient in the City Hospital for Infectious Diseases. Any visitors had to stand outside a perimeter fence and the patients stood on the other side keeping to a marked line: 'in peril you must not cross the line'.

'There was a time when once a patient passed inside its gates all physical contact with the outside world was broken'. (Joseph McKiernan, writing in the *Evening Chronicle*, 27 September 1963.)

Despite the high level of nursing care in Newcastle's hospitals and the stringent attempts to prevent the spread of infectious diseases, the standards of health among the poor children of the city remained alarmingly low.

The Annual Report of the Newcastle Dispensary in 1931 indicated grave concern for the physical well-being of the 'poorer classes'. The concern caused by this report led to a study by Dr James Spence concluding that the malnutrition and ill health that he found in working-class children were due to the conditions in which they lived. The experiences of the Second World War endorsed these conclusions. Evacuation illustrated the glaring differences in the health and well-being of children from different social groups. Government control over food supplies and the introduction of rationing brought about radical changes in diet. Vitamins and food supplements were provided for pregnant and nursing mothers and young children.

A children's ward, the Royal Victoria Infirmary, shortly after opening.

When the RVI was built, 1900-6, much care was taken with the decoration of the children's wards. There were 61 Doulton tile pictures, mostly of nursery rhymes, on the walls. The Lady Mayoress and her friends paid for them.

The Christmas tree and Father Christmas talking to a young patient indicate further the concern to create a reassuring atmosphere for the children.

On a local level, a landmark was the creation in 1943 of the Chair of Child Health, King's College, Newcastle upon Tyne. Dr Spence was appointed as the first full-time Professor of Paediatrics in England. His concern for families was paramount; he pioneered the practice of admitting mothers into hospital to comfort and help with the nursing of their sick children. Today, in the Royal Victoria Infirmary, children are treated in the Sir James Spence Institute of Child Health.

When the war was ending in 1945, the British electorate voted for the Labour Party led by Clement Attlee. The new government brought in major reforms, creating the National Health Service and the Welfare State: 'A shield for every man, woman and child in the country against the ravages of poverty and adversity'. (James Griffiths, Minister of National Insurance, 1945-50.)

In Newcastle a detailed survey of 'a thousand families' began. A record was kept of the health and schooling of approximately 1,000 children, who became known as the 'red spot children', born in 1947 until their fifteenth birthdays. The survey concluded: 'The decline in deaths during infancy and early childhood has continued and the figure is the lowest ever recorded. Death in childhood is rare and parents can expect to rear children who survive the first week of life.' (F.J.W. Miller, *Growing up in Newcastle upon Tyne*, Oxford, 1960.)

The survey gave the reasons for these conclusions as full employment, improving standards of education, the spread of information on the care of children through child welfare clinics and advances in medicine. Writing this book in 1996 it is difficult to be as optimistic. It is clear, however, that the future of Newcastle lies with its children and that everything must be done in education and child health care to ensure the future is bright.

A short bibliography

Barton, Arthur, *Two Lamps in our Street*, Hutchinson, 1967.

Callaghan, Tom, *Those were the Days*, Newcastle City Libraries, 1992.

Common, Jack, *Kiddar's Luck*, Bloodaxe Books, 1986.

Forster, Eric, *The Pit Children*, Frank Graham, 1978.

Hind, Joe, *A Shieldfield Childhood*, Newcastle City Libraries, 1994.

McCord, Norman, *North East England. The Region's Development*, 1760-1960, Batsford Academic, 1979.

Miller, F.J.W., Court, S.D.M., Knox, E.G., Brandon, S., *The School Years in Newcastle upon Tyne*, Oxford, 1974.

Miller, F.J.W., Court, S.D.M., Walton, W.S., Knox, E.G., *Growing up in Newcastle upon Tyne*, Oxford, 1960.

Peacock, Basil, *A Newcastle Boyhood, 1898-1914*, Newcastle City Libraries and London Borough of Sutton, 1986.

Spence, J.C., Walton, W.S., Miller, F.J.W., Court, S.D.M., *A Thousand Families in Newcastle upon Tyne*, Oxford, 1954.